Praise for *The Hop Grower's Handbook*

"If you want to grow hops, read this book through; then put it on your nightstand and read it every night to remind yourself of what's coming next. It's a seriously informative and surprisingly engaging book, full of resources and wise advice about all phases of growing, harvesting, and selling the buds that flavor the brew."

—JOAN DYE GUSSOW, author of *Growing, Older* and *This Organic Life*

"While driving on rural roads in the northeastern United States, if you look carefully you may notice wild hop vines twining their way up utility poles and signposts, feral reminders of a once-thriving hop industry and a time when brewing and the production of the raw materials used in beermaking were both local endeavors. And while craft brewers helped launch the local food movement over three decades ago, the local production of brewing ingredients has lagged behind. *The Hop Grower's Handbook*, packed with research and practical advice, is an invaluable tool for reuniting regional brewers with regional growers. This delightful and useful book should be part of any brewer's or small-scale farmer's essential library."

—PETER EGELSTON, founder and president,
Smuttynose Brewing Company

"I have spent the last several years working with an eclectic group of new hop growers, including Ten Eyck and Gehring, in New York and other eastern states. Up until now I have had to tell them that there is no 'cookbook' for growing hops on a smaller, but commercial, scale. There is much information on the Internet—but most is anecdotal, unproven, and geared toward backyard home-brew enthusiasts. Now I can point people to this well-thought-out and informative publication. This is a great step forward for the new hop industry!"

—STEVE MILLER, hops specialist, Cornell University

"How exciting to begin to see the principles of healthy growing brought to the hop yard! This handbook richly shares wisdom on hop horticulture as well as trellising, harvesting, and drying methods. But Ten Eyck and Gehring properly take the discussion further—weaving in biodiversity, disease acumen, and compost-based nutrition. Local brews deserve local hops, creating yet another quintessential niche for a savvy grower."

—MICHAEL PHILLIPS, author of *The Holistic Orchard*
and coauthor of *The Herbalist's Way*

"Beer from here! In their quest to re-establish hop growing in the Northeast, Laura Ten Eyck and Dietrich Gehring have compiled a holistic and adventuresome volume. We all benefit from their curiosity and action-research tactics. The local beer and brewery scene is heating up fast, and demand for local ingredients will continue to grow. I urge young farmers to study up and get in on the action!"

—SEVERINE VON TSCHARNER FLEMING,
director, Greenhorns

"Given the recent explosion of hop production in the Northeast, new growers are frantically looking for information. So, welcome to *The Hop Grower's Handbook*—the only book covering practical hops production in the region. A timely and highly valuable resource for growers!"
—HEATHER DARBY, agronomy specialist, University of Vermont

"Hop to it—but not so fast! It's best to learn from the trials and travails of the pioneers who have somehow busted through what recent generations have considered an impasse: growing the hops for truly local beer and reviving a commercial hop industry in parts of the United States that a century ago caved to the success of growers in other, far-flung regions. A how-to guide without a narrative of successes, failures, and subsequent innovations is a recipe for disaster, not to mention boredom. Laura Ten Eyck and Dietrich Gehring lived and share the compelling story of the recent resurgence of hop production in the Northeast—a high-flying grassroots movement bringing together hopophiles, dusted-off texts of a bygone era, terroir aficionados, cutting-edge farmers, and creative extension agents. Their book elevates hops knowledge to a new level while making even the most far-reaching possibilities tenable. *Prost* to the plant ready to take over and offer yet another convivial contribution to our local food—and drink—renaissance! Hop, hop, hoorah!"
—PHILIP ACKERMAN-LEIST, author of *Rebuilding the Foodshed*

"*The Hop Grower's Handbook* is a fantastic source of information that will absolutely help revive the small-scale hop-growing industry of New York and other eastern states. Ten Eyck and Gehring give experienced and novice farmers alike a structured plan to efficiently start and maintain a hop yard at the most ideal scale for today's local hop markets. Farm brewers and microbreweries should be excited, as this book will surely help make locally grown hops more readily available in the coming years. The authors and their Indian Ladder Farmstead Brewery and Cidery have continued to stay ahead of the curve on hop growing, harvesting, and processing. The hard work and time put into this book benefits all of us in the small-scale brewing industry. As a Brooklyn-based brewer, I'm eager to watch the regrowth of a product with such heritage in the state of New York, and I look forward to making more great beers brewed with more quality, locally grown hops."

—MATT MONAHAN, cofounder and co-owner,
Other Half Brewing Company

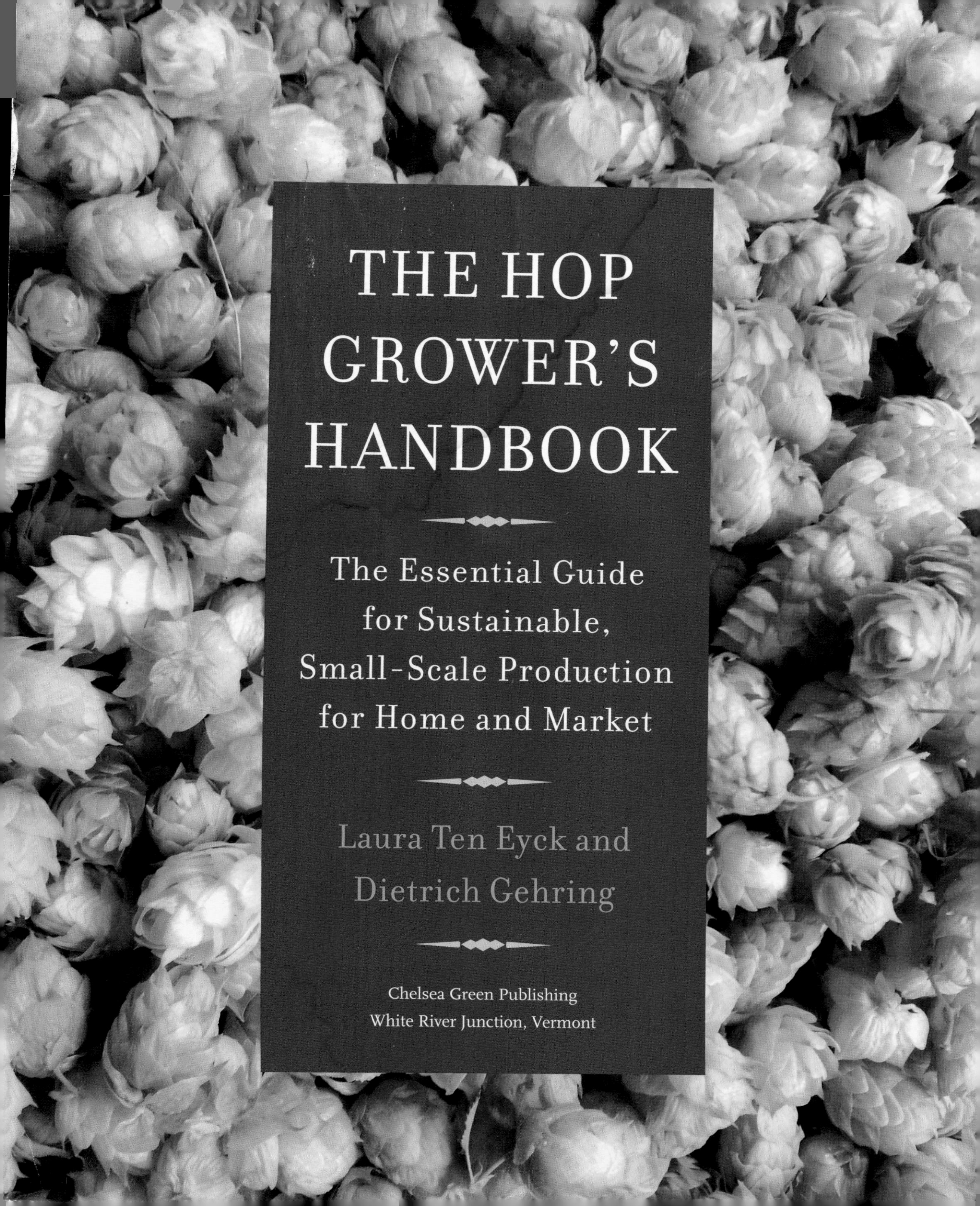

THE HOP GROWER'S HANDBOOK

The Essential Guide for Sustainable, Small-Scale Production for Home and Market

Laura Ten Eyck and Dietrich Gehring

Chelsea Green Publishing
White River Junction, Vermont

CONTENTS

Introduction

It was 7:15 a.m. on a hot and muggy Saturday morning in mid-July when I woke. Despite the early hour my husband, nicknamed Dieter, was gone from the bed and it was already pushing 80 degrees. Here in upstate New York we were about to suffer through the sixth day of an unbearable heat wave blanketing the Northeast. Last night we had sat on our deck overlooking the pond, mesmerized by heat lightning flashing across the night sky. The peach-tinted waves of light illuminated the scalloped edges of a brigade of dark thunderheads massing on the northern horizon for a storm that never came. Prior to the July heat wave we had endured a month of heavy rain. The combination of the prolonged deluge followed by the period of intense heat had resulted in massive swarms of insects, and on the deck we were soon enveloped in a cloud of mosquitoes that eventually drove us into the house. To beat the heat Dieter had taken to getting up early to do farm chores, patrol our newly established hop yards for the latest outbreaks of insects and disease, and cut down the rampant weeds flourishing in the heat and moisture.

Helderberg hops growing in front of our barn, with a view of our pilot hop yard the first year it was planted.

A panorama image of Indian Ladder Farms beneath the Helderberg Escarpment.

recruiting friends to work the Newman's beer truck at festivals such as Albany's Pinksterfest, we drank a lot of it. Although the Wm. S. Newman Brewing Company eventually went out of business, the craft beer industry took off. Dieter got into home brewing, led the formation of a local beer club that meets monthly to try new beers, started a beer blog that became very popular, and began growing hops in the garden—and up the side of the house.

But before beer there was farming, and in hindsight it was inevitable that these two occupations and preoccupations would merge in our lives. Both our families have roots in dairy farming. Dieter's family had a dairy farm in New York's Mohawk Valley, and he spent many summers of his youth living and working there. My family's farm, on which we live today, was originally also a dairy farm. It was started by my great-grandfather, Peter Ten Eyck, a businessman, politician, and farmer who in 1915 purchased a tract of land made up of five individual farms in western Albany County beneath the limestone cliffs of the Helderberg Escarpment. There he created a single farm and named it Indian Ladder Farms, after a Native American trail that once scaled the cliff face. The farm started out as an orchard and dairy, then turned to raising beef cattle, and eventually converted

Dieter's mother as a child, standing on her father's shoulders on Matis Farm in St. Johnsville, New York.
Photograph Courtesy of the Matis Family

Bins of apples on Indian Ladder Farms, in the Ten Eyck family since 1915.

entirely to orchard. Today the farm, operated by my father, who is also named Peter Ten Eyck, remains primarily an apple orchard, with a sizable retail farm market and pick-your-own business. As I write, he is in the process of retiring and turning over management of the business to my brother, another Peter Ten Eyck, and me.

After we moved back to the farm, Dieter and I experimented with various types of agriculture. We grew specialty vegetables for restaurants, kept a flock of sheep for meat and wool, raised chickens for eggs and meat, and created a small herd of dairy goats. Today we continue to produce much of our own food, gardening extensively and raising small livestock. I have worked on my family's farm ever since I was a child. As an adult, giving up on my not-necessarily-lucrative career as a freelance journalist and newspaper reporter, I served a long stint helping my father run the farm's retail operation before launching a new career in farmland conservation.

Today I work for the New York State office of the national farmland conservation organization American Farmland Trust. Dieter, a photographer and photo editor by trade, also helped out on Indian Ladder Farms over the years. A while ago my father, in his seventies, gave us the house we live in along with 60 acres (24.3 hectares) of land. Twenty acres of the land we now own are still cultivated by Indian Ladder Farms through an informal lease arrangement; however, the remainder of it is not. After taking ownership of the land we began to cogitate about what we should do with it. Right around that time New York State adopted new legislation called the New York State Farm Brewery Act, intended to stimulate economic development in the state by elevating beer making to the same level as the state's extremely successful vineyards and wineries. Brewing beer was now not only allowed but encouraged on farms as long as it was made with ingredients grown in New York State.

Needless to say, a lightbulb went on in Dieter's head. We began planting more hops and barley, started a hops test plot, and eventually put in a small-scale commercial hop yard. Today Dieter divides his time between photography and farming, and he and our business partner, Stuart Morris (a friend from our days at Beacon Hill Wine & Spirits) have launched the Indian Ladder Farmstead Brewery and Cidery. Like all Eastern hop farms, ours is new by agricultural standards. Nearly thirty years of growing hops for pleasure has given us insight into the plant, but having commercial aspirations and a full hop yard has brought many new lessons our way.

Which brings me back to the day we had our first Japanese beetle alarm. While we drank our coffee on the porch, Dieter called my father, who went to Cornell's School of Agriculture in the 1950s and majored in insects and how to kill them. Indian Ladder Farms is not an organic orchard, but my father is committed to using as few pesticides as he can. The farm operates under the terms of an environmental label called Eco Apples, which ensures its growers adhere to strict Integrated Pest Management principles and a low-spray program. Located in the middle of a nonorganic apple orchard, our hop farm will never be able to obtain organic certification—but our goal is to farm in an environmentally sustainable way. My father said he was also fighting Japanese beetles in the apple orchard and told Dieter he was going to have to spray the orchard with insecticide to kill them. Dieter decided to try another route—spraying the hop bines with neem oil, made from the crushed fruit and seeds of the neem tree, which grows in India. Neem oil contains azadirachtin, a natural pesticide approved for organic use. He also set up some Japanese beetle traps near the hop yards. By the next morning the traps were filled with thousands of Japanese beetles, and during his morning patrol Dieter noticed many of the beetles still clinging to the hop leaves were in fact dead. One problem down.

A couple of weeks later we went on our annual one-week pilgrimage to the beach. We expected the hops to come to maturity a week or ten days after our return. Dieter planned to spend the time in between

Our pilot hop yard before the trellis collapse. Note the trellis is beginning to sag under the weight of the hops.

our return and the harvest setting up the drying area in the barn. When we returned we found that a section of the hop trellis had collapsed beneath the weight of the bines. After using the tractor to try to pull the collapsed section back up and nearly taking down the entire hop yard and trellis in the process, we decided to cut the bines down even though all the hop cones weren't quite ready.

We spent the entire day on our deck with friends pulling the cones off the bines. We were amazed by how many flowers there were. It was just our second year, so we had not expected the bines to produce so many so soon—and we didn't have a mechanical harvester yet. I began to feel extremely stressed, but Dieter seemed to be perfectly content drinking beer and picking hops. How in the world were we going to harvest all of this by hand? But the more immediate question was the following: without the drying area set up in the barn, where were we going to dry several bushels of hop flowers? Picked hops must be spread out to dry immediately after harvest so they don't mold.

The section of collapsed trellis we encountered upon returning from vacation.

Hops are a soporific, and hop picking made me very sleepy. I went upstairs to lie down for a bit. I heard a lot of commotion in the living room but tried to ignore it. When I came back down I found all the furniture had been pushed to the edges of the room and a giant blue tarp had been spread across the entire floor. The tarp was covered with a layer of hops and surrounded by four box fans turned on high. Dieter and a couple of our male friends sat around drinking beer and waiting to see how mad I would be. Absurdly, a small child-size rake lay in the middle of the sea of hops. I had to laugh—because if I didn't I would have cried.

These were just some of the new challenges that reared up when we began to scale up our garden-variety hop hobby to a commercial venture. As a gardener, whether you are graduating from cultivating tomato plants in raised beds to field production, expanding a handful of backyard fruit trees to an orchard, or transitioning from a single hop bine growing in a corner of the garden to a hop yard, you will face new issues. Sure, you know how to take care of a tomato plant—but when you apply that knowledge to taking care of one hundred or one thousand plants, you find yourself on a different playing field entirely. Insects and disease can quickly run wild in even a small-scale monoculture planting. Meanwhile weeds will happily take advantage of the water and nutrients you are providing for your crop. And if you are successful your harvest will no longer be picked in a couple of hours on a pleasant afternoon. Because of the increased scale of a crop produced for market, controlling pests on a plant-by-plant basis, and weeding, you'll find cultivating and harvesting by hand may no longer be practical—especially if you are working on your own.

Today we are growing hops on a small-scale commercial basis. At our Indian Ladder Farmstead Brewery and Cidery, we are making hard cider, flavored with our own hops, from apples grown on Indian Ladder Farms—as well as beer made with the hops and barley that we grow along with some that we purchase from other farmers in New York State. The brewery has a tasting room where people

We tried to pull the trellis back up but it didn't work. Photograph by Laura Ten Eyck

Collecting an early harvest in the bucket loader. Photograph by Laura Ten Eyck

Our son, Wolfgang, carrying hop bines to the deck where we handpicked the cones. Photograph by Laura Ten Eyck

Laurie handpicking hops on the deck.

Bucket starting to fill with hops from collapsed trellis.

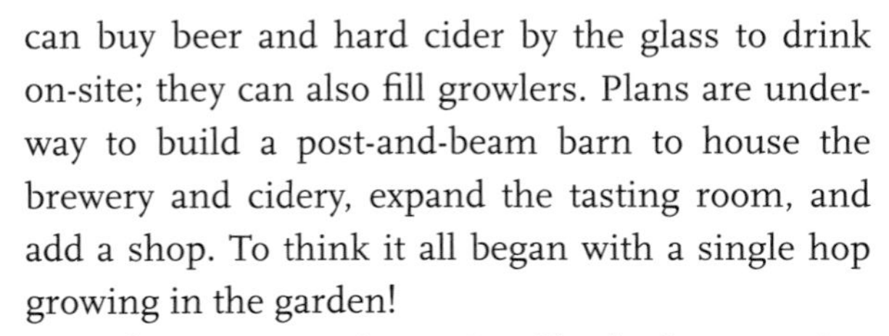

can buy beer and hard cider by the glass to drink on-site; they can also fill growlers. Plans are underway to build a post-and-beam barn to house the brewery and cidery, expand the tasting room, and add a shop. To think it all began with a single hop growing in the garden!

And we are not alone. Outside the hop-growing stronghold of the Northwest, productive hop yards have taken root in other states, including Colorado, Michigan, Wisconsin, Minnesota, and Ohio. In the eastern United States, hops are being grown from Maine to North Carolina. Whether they are homesteaders, home brewers, or in it to make money, people are growing hops and using those hops to brew local beer.

The Hop Yard in Gorham, Maine, got its start on a potato farm and is now supplying the craft beer scene in Portland, Maine. Addison Hop Farm supplies its organic hops to breweries and home brewers in its home state of Vermont, which has the largest number of breweries per capita of any state in the nation. The Hop Farm Brewing Company, in Pittsburgh, has its own hop yard outside the city and is working with regional farmers to increase hop acreage in western Pennsylvania. Old Dominion Hops Cooperative is a group of ninety-plus farmers producing local, sustainably grown hops to supply the craft beer industry in Maryland, Virginia, and North Carolina.

When we first started planting hops beyond the garden gate there were few resources to guide

Our first hop harvest on the living room floor.

Easterners interested in growing hops in the garden, on a small scale, or commercially. The number of East Coast craft brewers was rapidly expanding, and many were starting to clamor for locally grown hops. But all the books, research papers, and online resources available were primarily about hop production in the Northwest, which takes place at a vast scale under completely different climatic conditions. By 2009, under the direction of agronomist Heather Darby, the Hops Project at the University of Vermont had put in their experimental organic hop yard—and results from their variety trials and research into diseases and insect pests began appearing online. The Northeast Hop Alliance formed, and a couple of conferences were held in New York and Vermont. The year after we put in our pilot hop yard, Cornell University put in their own experimental hop yard; in 2014 the Northeast Hop Alliance and Cornell put out the *Cornell Integrated Hops Production Guide*.

We devoured all the material we could get our hands on. It was clear that lots and lots of great science had been done on hop nutritional needs, varieties, insects, and disease—but nowhere could we find self-contained, simple, step-by-step instructions for growing hops commercially (hop-yard construction, planting, tending, and harvesting and processing). To complicate matters further, whether it was about soils, bugs, disease, or the hops themselves, most of the language used was highly technical and difficult to understand. We've worked hard to learn what we have learned, and we are still learning. So we decided to create the book we wish we had when we started out, in the hopes that it will help others who want to grow hops, either for their

Dieter holding hop bine in the one-acre hop yard.
Photograph courtesy of John Carl D'Annibale/*Times Union*

Laurie checking out a sample from a brewer's cut in the Yakima Valley, Washington.

Wolfgang paying homage to hops at Crosby Hop Farm in the Willamette Valley, Oregon.

own use or for market and help fuel the renaissance of small-scale, sustainable hop growing in the East.

In the pages ahead, you'll gain a general understanding of the hop plant itself, its botany, its history, and its role in the brewing of beer. You'll also learn about hop varieties, how and where to grow and care for them, and what kind of infrastructure and equipment you will need at various scales—whether you are growing hops on a city balcony, in a backyard garden, in a miniature hop yard, or as the real thing. We also explain how to process the cones to ensure the highest quality for beer making, explain the basics of how beer is made, and talk about the role of hops in brewing beer. As a grower, the more you know about how hops are used by brewers, the better. At the end of the book, friends who are commercial brewers and home brewers share their recipes for beer they have made with our hops. One thing we have found is that, like drinking beer, growing hops is a social activity that brings together friends and community.

PART I
REDISCOVERING HOPS

CHAPTER 1

Hop History

Today, it's nearly impossible to think about hops without thinking about beer. But humans were using hops for millions of years before someone—no one knows who—had a eureka moment and added hops to his brew. Paleobotanists have found evidence of fossilized hops pollen—along with pollen from its close relative in the Cannabaceae family, marijuana—at numerous archaeological sites. In Paleolithic times, humans who still chipped stone for spear points used the wild hop varieties that grew in what are now North America, Europe, and Asia as one of their many medicinal plants. Throughout the ages that followed, hops were used to ward off plague and treat everything from jaundice to toothaches. The bitter herb is still valued for its healing properties.

The earliest known written reference to hops (which mentions the plant as food, not medicine) is found in *Naturalis Historia*, published in 77–79 AD by Roman author Gaius Plinius Secundus—popularly known as Pliny the Elder. The thirty-seven books included in this wide-ranging encyclopedia contain Pliny's thoughts about and understanding (and sometimes misunderstanding) of natural history. Readers encounter passages on "the extreme smallness of insects" and "the belly: animals which have no belly. Which are the only animals that vomit." It's not until Chapter 50 of Book 21 that we stumble upon the wild foods knowledge of the times. Here, Pliny refers to a plant called *Lupus salactarius*, or "wolf of the willows," widely believed to be a reference to hops and named for the plant's rampant

Pliny the Elder provided the first written account of hops in *Naturalis Historia*, published in ancient Rome. Photograph Courtesy of Paul D. Stewart/Science Photo Library

growth and tendency to climb and engulf any plant or tree that happens to be growing nearby. Although he doesn't go into preparation techniques, it is likely

From Ancient Remedies to Modern Medicine

Today hops are used far more widely in brew kettles than in medicine chests, but herbalists never really gave up on the plant, using it for anxiety, sleep disorders, ADHD, irritability, and indigestion; to increase appetite and urine flow, stimulate breast milk flow, assist cancer patients, and heal leg ulcers; and for numerous other uses. Today, scientists too are rediscovering the medicinal uses of hops, and especially the following ones.

KEEPING IT CLEAN

Hops' antibacterial properties were a major reason the herb became popular for beer. As an antibacterial agent, hops kill bacteria in beer and allow it to

In addition to flavoring beer, hops have many medicinal purposes and have been used throughout the ages as an antibacterial agent and soporific, among other uses.

that cooks in Pliny's day sautéed or steamed the plant's tender shoots—a practice that continues today, primarily in Europe, where the spring shoots are referred to as "hop asparagus" (for a recipe, see Chapter 6). The Roman naturalist, philosopher, and military man was never able to expand upon this early record of hop use, or anything else for that matter. On the verge of the publication of *Naturalis Historia*, Pliny, an inquisitive man of science to the end, met his death after traveling to Pompeii to more closely inspect an unusually large cloud mass forming over the volcano Mount Vesuvius.

At this time, wild hops may have played a very minor role in beer making as one of a varied cast of

have a longer shelf life. Hops have long been used as poultices on wounds, presumably for the purpose of preventing infection. Research has attributed the hops' ability to fight bacteria to the compound humulone, found in the hop resin.

OF MENOPAUSE AND MEN

Like soy plants, hops contain phytoestrogen, a plant-produced compound similar to estrogen. Hops have long been used to treat "female complaints," and research is now being done on whether the phytoestrogen produced by hops can replace pharmaceutical estrogen to alleviate symptoms of menopause. It's not surprising, then, that hop leaves were once fed to dairy cows to increase their milk production, and many nursing mothers have been told to drink a beer to help with the release of breast milk.

A twist on this is that hops have been accused of acting as an anaphrodisiac (reducer of sexual desire) in men. It has been rumored that women and children made up the bulk of hop-picking crews in the nineteenth century because picking hops caused men to grow "beer breasts."

A somewhat obscure slang term for impotence, "brewer's droop," is an old-fashioned term applied to men afflicted with the condition who worked extensively with hops and beer, such as brewers and tavern keepers. Brewer's droop was attributed to their excessive exposure to hops. Today impotence is acknowledged by modern medicine to be a symptom more frequently associated with excessive alcohol consumption. But a connection between "brewer's droop" and the phytoestrogen in hops may in fact be real.

SLEEP TIGHT

Hops have long been known to function as a soporific. This is something I can personally attest to: when picking hops I become very sleepy. Pillows were often stuffed with dry hop cones to create a fragrant, sleep-inducing cushion for insomniacs, and hop tea is used in lieu of chamomile tea as a bedtime ritual. I recently spoke with a farmer who suffered from insomnia and treated herself with a homemade tincture of hops. To create the tincture she steeped dried hop flowers in high-proof alcohol. Technically an herbal tincture is one part herb to three parts alcohol. But she said she simply stuffed a jar with hop cones as full as she could, filled the jar with grain alcohol, and let it steep. A dropper full of the tincture before bed put her right to sleep. She's wasn't quite sure whether it was the hops or the alcohol that put her out, but she didn't care because it worked.

CANCER FIGHTER

Research has shown that antioxidants found in hop flavonoids may be more potent than those found in red wine and green tea. Antioxidants are believed to help the human body fight cancer. One of the flavonoids—xanthohumol, found only in hops—is proving to be extremely powerful. Due to a unique protective coating, it can survive even longer in the body than other known flavonoids.

herbs. Herbs were selected at the discretion of the brewer, each of whom had his (or often her) own closely held recipe. Herbal blends used in brewing were called *gruit* (pronounced "groot"). A blend of plants such as sweet gale, yarrow, and marsh rosemary was considered a standard—but the mixture could include a variety of other plants, such as juniper, ginger, caraway seed, anise seed, nutmeg, and cinnamon. Like hops, which would eventually take over as the primary brewing botanical, the various herbal blends provided flavoring, medicinal properties, and preservative action for beer. However, unlike hops, when fermented many of the herbs used in gruit enhanced the inebriating effects of the alcohol in beer with their own narcotic, aphrodisiac, and hallucinogenic properties, making beer brewed with gruit an extremely popular product.

Nonetheless hops had their uses, and although (like the other gruit herbs) they were primarily collected in the wild, people eventually began to grow them. Even though hops were competing with a lot of other herbs for time in the brew kettle, it is clear they were valued. References to exchanges of land on which wild hops grew plentifully appear in the eighth century. In 768 AD Pepin the Short, a man of note due to the fact that he was the father of Charlemagne, the King of the Franks, donated land to the Abbey of St. Denis, located near Paris. It was worth noting in the deed that an abundance of wild hops grew on the land.

Less than a century later, in 822 AD, came the first written reference to hops used in beer. Although it may appear that, considering all the brewing of beer and aphrodisiacal gruit, these were fairly lawless times, it seems that there were even more rules governing everyday life than there are now. Corbie Abbey was a Benedictine monastery in France. The abbey's rulebook specified that everyone who collected firewood and hops had to give a tenth of what they collected (a tithe) to the monastery porter—the monk who, instead of being quartered in the dormitory along with the other monks, had a room near the entry so that he could interact with the public on behalf of the monastery. The Corbie Abbey rule book of 822 stipulates that if the amount of hops given to the porter was not sufficient for him to make enough beer for himself it was okay for him to get his hops elsewhere. It is worth noting that these hops were being "collected," meaning they were likely being gathered from the wild and not yet being cultivated.

Later in the ninth century references to hops being cultivated in what is today Germany begin to appear in the written record. For a long time the English had a thing against using hops to brew beer, associating hopped beer with the Dutch merchants whom they disdained. As a result historians believe that hops were not used in brewing beer in England until the end of the medieval period. But archeological evidence of a much earlier trade in hops involving England sprang onto the scene in 1970, when workers digging a drainage ditch in the Graveney Marshes in North Kent, near the English Channel, discovered an extremely old sunken boat. Maritime archeologists have identified the wooden boat as an Anglo-Saxon clinker-built boat, a design featuring a hull of overlapping planks and capable of carrying up to 5 tons (4.5 metric tons) of cargo and four men. The boat dates back to the early tenth century. Its primary cargo appears to have been hops. Historians are somewhat confused as to why this boatload of hops sank in an English marsh thousands of years before hops were thought to have been used in any quantity in English beer. It is possible the hops were intended for somewhere other than the brew kettle. In those days, hops were used in medieval cures for everything from insomnia to infection; they were boiled for dyes; and their fiber was used to make both paper and rope.

By the twelfth century people were well aware that the antiseptic properties that made hops good medicine also made them a good preservative. Around 1150 the Benedictine abbess Hildegard of Bingen—the German nun, writer, visionary, and mystic also known as Saint Hildegard—mentioned hops in a medical book often referred to as *Physica*, which she herself called *Secret Powers Hidden in Natural Creatures for the Use of Human Beings by God*. Hildegard did not recommend hops for medicinal use. The hop, she wrote, "is not much use for a human being since it causes his melancholy to increase, gives him a sad mind, and makes his intestines heavy. Nevertheless its bitterness inhibits some spoilage in beverages to which it is added, making them last longe."

Indeed, the preservative function of the hop enabled brewers to produce beer in larger quantities and transport it greater distances, reaching a broader market. Cultivation of hops for brewing beer began

Saint Hildegard of Bingen, who was offered to the Catholic Church as a child and started seeing visions of luminous objects at the age of three, studied science and medicine and advised using hops to preserve beverages. Photograph of *Miniatur aus dem Rupertsberger Codex des Liber Scivias*, courtesy of Wikimedia Commons

John the Fearless established the chivalric Order of the Hop. The order's emblem was a wreath of hop vines around the Duchy of Burgundy coat of arms. Jean-Jacques Chifflet, *Lilium Francicum* (1658)

in Germany and slowly spread throughout Europe, arriving in England at the end of the Middle Ages.

However, beer brewed with hops was not necessarily welcomed everywhere. Whether for its enhanced inebriative effect or simply for its accustomed flavor, brewers continued to brew beer with gruit. There were also political reasons for the persistence of gruit. Herbs used in gruit were gathered from the wild. During feudal times the wild—meaning essentially all natural resources—was the property of the crown. The right to gather gruit herbs—"gruit rights"—was granted to nobles and monasteries in exchange for loyalty. The gruit rights were then doled out to those who brewed beer in exchange for a tax. Initially the use of cultivated hops in brewing was discouraged because the powers that be didn't want to lose out on this tax. Nobles flaunting their independence from the crown encouraged their peasants to grow hops. In 1409, in order to make himself popular with his hop-growing subjects in Flanders (a major beer-producing region) the Duke of Burgundy, known as John the Fearless, established the chivalric Order of the Hop—a fellowship of knights whose emblem was a wreath of hop vines around the Duchy of Burgundy coat of arms. Apparently establishing the Order of the Hop got the Duke of Burgundy the attention he desired. John the Fearless is commemorated today in the Belgian town of Poperinge's Hoppefeesten, a beer and hop festival and pageant held every three years.

Eventually the problem of lost income from gruit taxes was solved by simply imposing a tax on cultivated hops as well. Today gruit is making a comeback among home and craft brewers of an experimental bent. Some modern-day gruit enthusiasts consider the demise of gruit to be a conspiracy of sorts and view the substitution of hops for gruit as the original temperance movement and a precursor to the criminalization of marijuana—a plot engineered by Protestant reformists who preferred hops as a natural sedative for the masses over gruit, which, with its hallucinogenic and aphrodisiac qualities, agitated the populace.

Whether the result of a Protestant plot or its qualities as a practical preservative, by the sixteenth century hops became the primary botanical used in beer making and an important agricultural crop. In 1574 Reynolde Scot, a country gentleman in England, published *A Perfite Platforme of a Hoppe Garden*, the first book about growing hops to be written in English. The book includes chapters on every aspect of hop cultivation. Observing pests, Scot writes that "the hoppe that liketh not his entertainment will be much bitten with a little black flye who if she leave the leaf as full of holes as a nette, yet she seldome proceedeth to the utter destruction of the hoppes." Chapters of interest to the new hop grower include "of the good hoppe" and "of the unkindly hop," as well as "when and where to lay the dung," plus the intriguing and sometimes all-too-familiar chapter "of diverse men[']s follies."

of a Hoppe garden

pedition you passe through your Garden, conti-
nually paring away eche greene thing assone as
it appeareth, you shall doe well, with the same,
and the vppermoste moulde of your Garden
togither, to maintaine and encrease the substance
of your hylles, euen till they be almoste a yarde
highe.

In the firste yeare make not your hyll to
rathe, least in the doing therof you oppresse some
of those springes which woulde otherwise haue
appeared out of the grounde.

It shall not be amisse nowe and then to passe
through your Garden, hauing in eche hande a
forked wande, directing aright such Hoppes as

declyne from the Poales, but some in steade of
the sayde forked wandes, vse to stande vppon a
stoole, and doe it with their handes.

A Perfite Platforme of a Hoppe Garden, written in 1574 by country gentleman Reynolde Scot, is the first book written in English about how to grow hops. *A Perfite Platforme of a Hoppe Garden* held in the Collections at the Library of Congress

By the mid-seventeenth century European growers began to make note of particular hop varieties and the characteristics they possessed. These varieties were named for the region in which they were discovered and propagated. Brewers began to seek out specific varieties for the properties they added to the beer, acknowledging that particular types of hops used in varying quantities produced different flavors and aromas. Growers paid attention to what the brewers wanted and did their best to provide it. These growers also began to take a second look at the volunteer hops that periodically sprouted up from the ground as a result of chance pollination, realizing the potential these plants might hold for the development of new traits that could appeal to beer drinkers.

But it was not until the eighteenth century that the hop trade really hit its stride and hops were being grown, traded, and used for brewing beer throughout Great Britain and continental Europe. Author Daniel Defoe, best known for his novel about a certain desert island castaway named Robinson Crusoe, observed booming hop sales firsthand at Stourbridge Fair, which (located outside Cambridge, England) was the largest medieval fair in Europe. Defoe wrote about the fair in his matter-of-fact travelogue *A Tour Thro' the Whole Island of Great Britain,* which he published in three volumes between 1724 and 1727. According to Defoe, traders came in large numbers from across Great Britain to set up rows of tents in a cornfield along the banks of the River Cam. He described the fair as a fortified city complete with coffeehouses, taverns, and cookshops to which people traveled from far and wide to trade in everything from gold and glass to hats and haberdashery. However, Defoe notes, all this was "outdone by two articles which are the peculiars of this fair . . . these are WOOLL and HOPS."

Although hops were not grown in the region, apparently a great quantity of hops was consumed there for beer, resulting in a market large enough to influence commodity pricing. "There is scarce any price fix'd for hops in England, til they know how they sell at Sturbridge Fair," wrote Defoe. "The quantity that appears at the fair is indeed prodigious, and they, as it were, possess a large part of the field in which the fair is kept, to themselves."

Thanks to the preservative properties of hops, trade in beer expanded far beyond Europe, particularly to European colonies. Beer produced for export to colonies in the tropics such as the West Indies and the Indian subcontinent had to survive a long sea voyage at high temperatures. Beer making its way to India sloshed in barrels in the hull of a ship for the duration of a four-month voyage, rounding Africa's Cape of Good Hope and crossing the equator not once but twice at temperatures that for a good part of the trip exceeded 80 degrees Fahrenheit (26.7 degrees Celsius). Anyone who has driven around in a car all summer with a few forgotten bottles of beer in the trunk knows that beer does not fare well under such conditions.

In his *Dictionary of Chemistry*, published in 1821, Andrew Ure, a Scottish doctor and scientist, wrote, "It is well known that other things being equal, the liquor keeps in proportion to the quantity of hops." He advised brewers to add 1 to 1½ pounds (0.5 to 0.7 kilogram) of hops to a 32-gallon (121.1 liter) barrel of "fresh beer" but to add 4 pounds (1.8 kilograms) of hops to those barrels bound for India. Following such advice, brewers added at least twice as many hops to a product they advertised as "Pale Ale prepared for the East and West India climate," which eventually became the India Pale Ale still popular today.

When the Dutch first began to colonize North America in the 1600s they found wild hops growing in the woods. This native New World variety, which would later be classified as *Humulus lupulus* var. *lupuloides,* had its own unique genetics. Of course Europeans migrating to North America brought hops with them from their homelands and cultivated them throughout the American colonies for making beer. When these various European hops began cross-pollinating with the wild American hops, a whole new set of properties was unleashed. The first named variety of hop resulting from such commingling was Cluster. The Dutch colonists grew their own hops, as well as importing them from the Netherlands, but between the middle of the eighteenth century and the beginning of the nineteenth century Massachusetts took the lead in North American hop production. The epicenter was Middlesex County—particularly the town of Wilmington, located 16 miles (25.7 kilometers) from Boston—which in 1780 harvested 30,000 pounds (13,608 kilograms) of hops. Massachusetts hops were even exported to Europe. But New England production waned after the Civil War, and the baton was passed to New York.

In 1808, James Cooledge, a Massachusetts man from the town of Stow (in Middlesex County) relocated to Bouckville, a town in central New York that would later become famous for hosting the annual

Madison-Bouckville Antique Week, the largest antique show in New York State. When he moved, Cooledge brought along a bunch of hop rhizomes that he planted on his Bouckville farm. The hops thrived in the freshly turned, nutrient-rich soil. By 1816 Cooledge was shipping his hops down to New York City. By 1840 New York State was producing almost as much hops as all of New England. Forty years later New York hit its peak, producing 21 million pounds (9.5 million kilograms) of hops on about 40,000 acres (16,200 hectares)—80 percent of the hops in the nation.

The new epicenter of hop production was Otsego County—home to James Fenimore Cooper, author of the *Leatherstocking Tales*, including *The Last of the Mohicans*. In 1880 Otsego County alone produced nearly 4.5 million pounds (2 million kilograms) of hops. The famous author's grandson, also James Fenimore Cooper, was a young man in Cooperstown during the height of New York's hop industry. "Those were the days when the 'hop was king' and the whole countryside was one great big hop yard and beautiful," wrote Cooper in a series of articles for the local newspaper entitled "Reminiscences of Mid-Victorian Cooperstown." "The air of the county was redolent of curing hops from thousands of hop kilns, burning all night. Then when the hops were baled and the pickers paid off and sent home, the town, probably the greatest hop market in America, filled up with buyers from all over the country. Truly it was, commercially, the golden age of the village never to be forgotten by those who lived through it."

Hop pickers in New York's Cherry Valley in Otsego County often traveled to the hop yards on trains from cities, their fare paid by the growers, who also provided room and board. For many, the hop harvest was a working vacation, picking hops by day and partying by night. Photograph courtesy of Timothy J. Albright

Over time, as with most monoculture, problems with disease and insects began to take their toll on New York's hop yards. These problems were greatly exacerbated by the Northeast's abundant rainfall and humid summers. By 1890 New York's acreage had dropped to 36,670 (14,840 hectares), but it was still by far the largest producer in the country. At this time commercial hop acreage in New England had declined to practically nothing. In Vermont there were 81 acres (32.8 hectares); Maine had 37 acres (15 hectares); New Hampshire, 15 (6.1 hectares); and Massachusetts, only 2 acres (0.8 hectares). Rhode Island, Connecticut, and New Jersey grew no hops at all.

Hop production in the Northwest had been started by New Yorkers such as Charles Carpenter, who, fleeing hop disease and insect outbreaks in 1872, dug up a bunch of hop rhizomes from his father's farm in Clinton County and headed west. He settled in Washington State's Yakima Valley, named for the resident Native American tribe. Happy to escape New York's rain and humidity, he planted his rhizomes in the high desert, watering his hops with the ice-cold snowmelt from the Cascade Mountains brought into the valley by the Yakima River. Others followed. By the turn of the century, the hop production in the Northwest was on the rise, with Washington growing 5,113 acres (2,069 hectares) of hops in 1890; California 3,970 acres (1,608 hectares); and Oregon 3,130 acres (1,267 hectares). In 2013 hop growers in Washington State harvested 27,062 acres (10,952 hectares) of hops.

For today's beginning hop grower in the Northeast—lacking knowledge and equipment, laboring over an acre or so of land much as farmers did a

Looking down from a hilltop vineyard on hundreds and hundreds of acres of Yakima Valley hop yards. Hop yards in the Yakima Valley are called ranches, average 450 acres (182.1 hectares) in size, and are family operated.

hundred years ago—it is difficult to comprehend the sheer scale of commercial hop production in places such as the Yakima Valley, where optimistic pioneers pushing westward, like Carpenter, settled and buried their rhizomes in the soil. In an effort to grasp that scale, Dieter and I made a pilgrimage to the Yakima Valley, located in south central Washington. There a big blue sky and rolling brown hills dominate 6,000 square miles (15,540 square kilometers) of high desert. The hills go by names such as Horse Heaven and Rattlesnake. The only green to be seen lies along the banks of the shimmering Yakima River, which briskly threads its way through the desert. The river's water is diverted by a system of dams and canals to irrigate orchards, vineyards, and thousands of acres of hops that materialize like mirages in the arid landscape. Amazingly, 80 percent of the hops produced in the United States—and 30 percent of the hops produced worldwide—are grown in this valley today.

But back in 1920 the enactment of Prohibition in the United States derailed hop production from coast to coast. Then in 1927 a lethal wave of the fungus downy mildew swept through New York's remaining hop yards, putting the final nail in the coffin of New York hop production. Two years later the passage of the Jones Act made it a felony to brew beer at home and people even lost interest in growing backyard hops. Despite an attack of downy mildew in the Northwest in 1930, hop production there held on long enough to recover with the repeal of Prohibition in 1933. The industry developed fungicides to control the still pervasive downy mildew and insecticides to battle the bugs infesting hop yards and deployed them in Washington and Oregon, where hop production would go on to dominate the world market.

Meanwhile European hop growers struggled with their own downy mildew outbreak, which circled the globe in the 1920s—not to mention two world wars. By the 1950s the northwestern United States was producing half of the hops worldwide. But American hops were unpopular in Europe. Even in colonial times Europeans disdained the characteristics that hops coming from North America lent to beer, and their dislike has continued into modern times. Only now, as American craft brewing infiltrates the global market, are some Europeans beginning to develop a taste for America's strong-flavored hops. Voices from abroad have described American hops as "rank," "piney," and even "catty." In 1882 a writer for the *Edinburgh Review* described American hops as having a "course, rank flavor and smell from the soil in which they grow which no management however careful has hitherto succeeded in neutralizing." Little did anyone know at the time that it would be these "undesirable" features of American hops that would lead to the wildly successful modern American craft beer movement, with its uber-hopped beers bearing names such as Heady Topper and Hop Head earning cult followers.

Nor do most of today's craft beer enthusiasts know just how much they owe to a man who, early on, became intrigued by the hop farmer's losing battle against disease and insects. In 1906 the interest of Ernest Salmon, a mycologist at England's Wye College, was piqued by the susceptibility of hops to fungal infections such as downy mildew. Salmon began to study the hop plant and experiment with breeding varieties with natural resistance to the disease. In search of ways to boost the defenses of long-cultivated European varieties, he had the radical notion of taking the hop back to its wild roots. Acknowledging the difficulty of finding a purely wild hop in Europe, where the plant had been cultivated for over a millennium, he looked across the pond to North America, a very big place where people had only been growing hops for a mere couple of centuries.

Salmon found a collaborator in Canada's national horticulturalist, Professor W. T. Macoun, who in 1916 obligingly sent Salmon a cutting snipped from one of the many hop plants growing wild along a shady stream near his research lab in Morden, Manitoba, just north of the Minnesota border. Macoun (yes, also a fruit grower, the professor is the namesake of the Macoun apple) enclosed a note with the sample: "The wild hops grow along a creek which flows through town. Old residents in Morden assure me there has never been any introduction of cultivated hops in this area."

Salmon propagated a plant from this cutting and christened it Wild Manitoba BB1. In Wye's research hop yard, BB1 parented varieties that had resistance to disease, but it came with a price. The offspring of BB1 possessed that intense smell reminiscent of cat urine that the Europeans found so distasteful. Salmon politely referred to the catty aroma as "that American nose" and set about attempting to tame BB1 through a refined breeding program that eventually produced Brewer's Gold. The extreme "hoppiness" of the resulting varieties were due to the high alpha acid content that is so sought after by modern craft brewers. All of the high-alpha varieties in existence today originate from Salmon's work with BB1.

After the repeal of Prohibition in 1933, breweries in the United States staged a brief comeback: 756 were brought back by 1934, only to undergo a massive consolidation of the industry. By 1950 there were only 407 breweries in operation and in 1961 only 230. By 1983, 6 breweries controlled 92 percent of US beer production.

Some American soldiers overseas during World War II had developed a taste for European beer and when they returned home found American beer lacking in flavor. Empowered by childhood memories of their fathers brewing beer in the family bathtub during Prohibition and a wartime resourcefulness, they set out to brew their own beer in an

Cornell University's Agricultural Experiment Station—located in Geneva, New York, and known for its long-term work with grape and apple varieties—installed a hop yard to trial hop varieties for production in the Northeast.

effort to make something better than what was available commercially. Home winemakers were way ahead of them, and when home brewing was eventually legalized shops that sold supplies to winemakers expanded to include home brewing supplies. Eventually things started getting serious, and some of the people who started out home brewing began producing beer on a commercial scale on the West Coast. The craft beer movement was born. Over time, the movement's followers developed a taste for intensely "hop-forward" beer, and the high-alpha hop varieties first developed by Salmon moved onto center stage.

Initially the craft breweries clustered in California and the Pacific Northwest. Eventually the enormously fast-growing industry spread across the country. According to the Brewers Association there were 2,768 craft breweries in the United States in 2013. Over 450 of these craft breweries were located in the Northeast. Yet regardless of their location the region's craft brewers continued to source the majority of their hops from the large hop yards in the Northwest.

Then along came the local foods movement and suddenly many people became intensely focused on issues like food miles and terroir. In 2007 the newly coined word *locavore* was awarded "word of the year" status by Oxford University Press. People wanted locally made beer, and they bought it, creating a boon for small local breweries. Eventually brewers wanted to take things a step further and explore making beer with locally grown ingredients, but they faced a big problem: hops, one of beer's most critical ingredients, were no longer grown anywhere but the Northwest. This was fine for West Coast brewers,

The University of Vermont started the Hops Project in 2009, experimenting with organic hop production on the university's research farm in Alburgh, Vermont.

and they capitalized on their ability to make a local product, holding hop festivals and featuring the farms where the hops were grown in their marketing.

Farmers in the Northeast, where there is a density of craft breweries and a history of growing hops, began to be interested. The University of Vermont (UVM) started the Hops Project to help farmers grow hops organically. They installed a hop yard on Borderview Research Farm in Alburgh, Vermont, and began to conduct field trials. Cornell University hired a hop specialist to help hop farmers get started and put in a trial hop yard at their New York State Agricultural Experiment Station in Geneva, New York. In 2011 new hop growers and the universities collaborated to form an organization called the Northeast Hop Alliance. In 2012 New York State really got the ball rolling by offering a farm brewery license, encouraging farmers and brewers to brew beer with ingredients grown in New York State.

The effort is starting to pay off. In 2010 there had been 15 acres (6.1 hectares) of hop yards in New York and even less in New England. By 2014 there were an estimated 250 acres (101.2 hectares) of hops growing in New York, and approximately 50 acres (20.2 hectares) in New England, with half of those located in Vermont. The craft beer movement and the local foods movement have converged, and the resulting renaissance in hop farming means farmers in the East are going back to the drawing board, relearning a form of agriculture that hasn't been practiced in a century.

As I write today, the hop-growing movement in the East is made up of a loose network of researchers and scattered farmers, brewers, and beer aficionados spread from New England and New York to Virginia and South Carolina with a lot of energy and enthusiasm, as well as a lot to learn. More and more farmers interested in joining the movement are seeking ways to start small-scale commercial operations. And more and more beer enthusiasts are seeking guidance on growing a few plants in their backyards for home brewing.

For those with commercial aspirations, major stumbling blocks are long-term access to farmland and the incredibly high costs of installing a hop yard. Another hurdle is the high price and scarcity of small to midsize hop-harvesting equipment and drying kilns appropriate for farms between 1 and 20 acres (0.4 and 8.1 hectares) in size. Beyond all this is the challenge of growing hops commercially in a wet, humid environment that fosters the very fungal diseases to which hops are most prone. Land grant universities running agricultural Extension services, such as UVM and Cornell, are providing research and support, but funding remains a mere fraction of what is provided for established industries such as dairy and apples. Growers and hop scientists in the Northwest are curious and supportive, but deeply familiar with the challenges involved, and they remain skeptical as to whether or not the Eastern growers can pull it off.

Eastern craft brewers like the idea of locally grown hops, but accustomed to getting any kind of hops they want at any time from the Northwest and around the world, it is not yet clear whether or not they will back the growers up as they face their many challenges. As with so many things in the United States the ultimate success of the enterprise will be decided by the consumer. If our local beer drinkers insist on beer made with locally grown ingredients and are willing to pay for this product, it is likely the hop industry in the East will succeed.

And because hops production in the eastern states is most likely to exist on a vastly smaller scale than in the Northwest, the region, with the right tools and practices, has a better chance of making sustainable production a norm, not an exception—meaning that the pioneer farmers bringing hops back in the region have much to share with small-scale market hops producers elsewhere.

The flowers of the hop bine, called hop cones, are rich in lupulin, a sticky resin that contains the acids and essential oils that flavor beer.

CHAPTER 2

The Hop Plant

With heavy bines towering high above you, being in a mature hop yard in late summer is somewhat like being in a very orderly forest. The bines wrapping around their coir are as thick and strong as cables. The mass of broad, deep green leaves soaks up the sunlight, pumping its energy into the formation of the flowers. Belowground, the plants' robust system of roots and rhizomes is essentially as large as what grows above, drawing up nutrients and water from the soil to nourish the massive amount of vegetation overhead. It amazes me to think of how effectively people have harnessed the incredible natural energy and characteristics of the hop—all to harvest the nuanced balance of acids

Maturing hop bines in the pilot hop yard.

hop plants from existing rhizomes rather than growing male plants for pollination. There are a few reasons for this. The first, and maybe the most compelling, reason is that the hop flowers produced by male plants, which don't even look like the cone-shaped flowers produced by the female plants, do not have what it takes to make good beer (more on this later). Second, hop growers do not want their female hop plants to be pollinated and go to seed. Seeds get mixed in with the hop flowers and produce added weight, which brewers, who buy hop flowers by the pound, do not want to pay for—especially since the seeds add nothing to the brewing process. In fact some brewers believe that hop seeds lend a bad flavor—something like rancid butter—to the beer. And finally, male hop plants produce a ton of pollen that, because of their high elevation, is carried on the wind to female hop plants far and wide.

In fact, before people started propagating hops from rhizomes, hop farmers would need to keep only one male plant per one hundred female plants in their hop yards. This prolific pollination effort on the part of the male hop plant results in undesired genetic crosses that can sprout up anywhere in the hop yard—even nestled within the established crown of a cultivated variety, its shoots indiscernible to the hop grower. Hops are heterozygous, meaning that a seed produced by a hop plant of a particular variety is unlikely to grow into a plant of the same variety as its parent. For this reason, when putting in new plantings people relied on hops propagated from rhizomes as it was the only way to ensure you were getting the variety you wanted. When Europeans arrived in North America, it is likely they did not bring hop seeds but instead brought plant cuttings and rhizomes.

All this is why only growers who are working on developing new varieties of hops keep male plants, and these are kept in isolation from commercial hop yards. If you have wild hop plants growing in the vicinity of your hop yard, keep an eye out for male plants. It is a good idea to remove them so they do not cross-pollinate with the female hops you are

The male plant produces flowers very different from the female plant in appearance. Because their flowers do not contain lupulin, the crucial element in hops for making beer, male plants are not welcome in commercial hop yards.

Female hop plants produce hop cones that contain lupulin. Most hop plantings contain only female plants, which are reproduced by propagating hop rhizomes.

Left to reproduce naturally, occasionally a hermaphrodite hop plant will occur that produces both male flowers and female hop cones.

cultivating. However, instead of completely destroying the plant, you may want to consider propagating a rhizome or cutting a sufficient distance from your hop yard to avoid cross-pollination in order to preserve the wild plant's genetics.

Hops' Seasonal Growth Cycle

Hops grow big, and they grow fast. To achieve this level of growth they need an incredible amount of energy from the sun, nutrients from the soil, and lots of water. In Chapter 6 we will track the hop as it grows and give detailed instructions for cultivation at each stage, but first let's take an introductory look at this incredible process.

In early spring the rhizomes in the plant's crown begin to send up a lot of shoots. It is up to the hop grower to prune these shoots back, selecting only a small number of the strongest shoots to begin to train to grow up the support system—whether that's

Hops are a vigorous, fast-growing plant that under the right conditions can grow as much as 1 foot in a single day.

proprietary and hail from the United States—but some are also coming out of New Zealand.

BETA ACIDS

Beta acids include the acids lupulone, colupulone, and adlupulone and are also responsible for bitterness in hops. Unlike alpha acids, beta acids release their bitterness primarily through the process of oxidation rather than isomerization, meaning that the acids break down and release bitterness when exposed to oxygen over time. The bitterness of beta acids emerges during brewing processes that involve air circulation, such as dry-hopping, or take place over extended periods of time, such as long-term fermentation. Beta acids can make up zero to 10 percent of the hop. Iso alpha acids tend to lose their power with age. This is where the beta acids come in, picking up the slack when the bittering power of the iso alpha acid begins to wane. It is the combination of alpha and beta that are important to the bitterness of the beer, and that is why brewers often look at the different levels as a ratio instead of a percentage.

During hop harvest the bines are normally cut and taken to a central location for harvesting. A section of the bine growing out of the ground about three feet in length is left behind. The remaining bine continues to photosynthesize, producing energy that is stored in the root system for the following spring. In the event a hop bine is not cut, as in the case of a wild hop, the additional vegetation will produce even more energy to feed the root system. The unharvested hop cones will simply brown and dry as they age. Once hard frost occurs, the vegetation aboveground dies back and the hop crown goes into a dormant state for the winter months.

Hop Varieties

Standing with a group of farmers, brewers, and plant scientists in a newly planted hop field in Geneva, New York, on a blustery August day, Dieter and I listened to Steve Miller—the one and only New York State Hops Specialist working for Cornell Cooperative Extension—answer a question that arises often: How should hop growers on modern-day, start-up hop farms in the East decide what varieties to plant? A long-time extension staffer focused on vegetable production Miller shifted gears in 2011 to become the extension's hop specialist, helping New York farmers expand the hops-growing renaissance playing out in the state, where the number of acres in hops production has grown significantly in the past ten years. But like others engaged in the East's effort to meet the demand for local hops, he knows the answer to this question is not easy. There are some varieties considered safer than others, but it will take experimentation by growers and scientists alike to really refine the options. Like the test plot we were standing in at the New York State Agricultural Experiment Station—where hop plants just 1 foot (30.5 centimeters) high were wrapped around strings suspended from a trellis system elevated by rows of newly erected 18-foot (5.5-meter) poles—hop farmers in the East are starting from scratch.

But these pioneers are not alone. Plant scientists at field research facilities such as Cornell University's New York State Agricultural Experiment Station and the University of Vermont's Hops Project at Borderview Research Farm are trying to figure out which of today's modern hop varieties are best suited to growing conditions in New York and New England. Their research is relatively new, and the fact that this perennial crop doesn't come into full production until its third year means their results are still a few years out. Meanwhile, the last best knowledge on the science of hop growing in this region is found in

Hop Oils, Aroma, and Flavor

Although bitterness plays a major role in beer, any beer aficionado will tell you it is only one member of a diverse cast. All the important character actors that give a beer its depth of flavor and aroma are found in the hop's essential oils. These essential oils are volatile, meaning that they turn from a solid or liquid into a gas. While bittering hops need to spend time in the brew kettle to release their bitterness, the aroma and flavor properties contained in the hop oil will completely vaporize during prolonged boiling and be lost. Brewers add aroma hops at the end of the boil to release the essential oil's properties without losing them. Hops varieties valued for their essential oils are referred to as "aroma hops." Because of the time at which they are added to the boiling wort—the liquid containing the sugars extracted from the malted barley—they are also called "finishing hops." Flavor imparted by essential oil is due to organic odor compounds such as humulene, myrcene, and caryophyllene. For example, humulene is associated with a woody/piney aroma, and myrcene is associated with an aroma described as green and resinous. Essential oil can be present in hops at a level between 0.5 and 4 percent. One interesting thing about hops grown in the East is that they seem to contain a higher level of essential oil than hops grown in the Northwest.

Cornell University's 1-acre (0.4-hectare) hop yard at the New York State Agricultural Experiment Station located in Geneva, New York, where researchers are conducting diverse hop variety trials and experimenting with methods of disease and insect control.

historical documents that people like Miller are dragging out of the archives.

There are over two hundred different varieties of hops, some newly developed and some from times gone by. Hop plants both wild and domestic contain a vast reservoir of genetic variables that people have experimented with over the years to create and mold hop varieties to meet their brewing needs. Some very old varieties are still in circulation, while others have fallen by the wayside. New varieties are being developed all the time to meet the evolving tastes of the consumer, produce a higher yield, and resist diseases and insect pests. The researchers who planted the nascent New York State Agricultural Experiment Station hop yard had selected thirty varieties to plant and monitor. Their goal was to see how well they would grow in the upstate New York climate: Can they fight off disease? Are they vulnerable to any particular kind of insect? What are the best means of controlling these threats?

Since powdery mildew and downy mildew pose the most serious threat to hop plants the world over, Miller,

TABLE 2.1. Hop Varieties Recommended for the Northeast

Variety	Brewing Usage	Aroma Characteristics	Yield	Maturity
Brewer's Gold	bittering	sharp, pungent, black current, fruit, spice	2,200 to 2,600 pounds per acre	late
Cascade	aroma	medium intense floral, citrus, grapefruit tones	1,800 to 2,200 pounds per acre	medium to medium late
Centennial	dual purpose	medium intense floral, citrus, lemon tones	1,430 to 1,700 pounds per acre	early to medium early
Fuggle	aroma	mild wood and fruit	1,000 to 1,400 pounds per acre	early
Liberty	aroma	mild yet spicy, subtle lemon, citrus tones	1,000 to 1,780 pounds per acre	medium early
Mount Hood	aroma	mild, herbal, somewhat pungent, spicy	1,250 to 1,960 pounds per acre	medium
Nugget	bittering	mild and pleasant, spicy, herbal tones	1,800 to 2,400 pounds per acre	medium late to late
Willamette	aroma	mild and pleasant, spicy, floral tones	1,700 to 2,200 pounds per acre	medium

Sources: Cornell University (for Northeast recommendations); Hopunion (for brewing usage and aroma characteristics); Oregon State University (for yield and maturity)

something new that is easier to grow. The hobby grower has the luxury of using her personal taste and curiosity as a guide, yet there are so many options that a scattershot approach to choosing varieties can quickly result in an overwhelming number of plants. In the United States a handful of popular varieties has emerged as reliable and multiuse.

PART II
PLANNING AND PLANTING

Because the hop plants are so large and heavy, they need support to reach their full size and produce the maximum harvest.

who wants to add the hop to one of your plant guilds? Are you a home brewer who wants to grow your own ingredients to make beer with? Maybe you are a homesteader who wants to grow hops for food, beer making, and medicine. Or are you a farmer who wants to sell hops commercially? If so, how much money are you trying to make growing hops? Will hops be your only crop or one element of a diversified farm? Your goals will define the size of the space you will need—a few square feet, the backyard, 1 acre, 10 acres, or 20 or more. Once you determine how many plants you need, you can determine how much space you need, and you can begin to look at your location options.

Hops are a great addition to a garden, but because they are a perennial and difficult to move, it is important to give them their own space and plenty of it.

Space Considerations for Home-Scale Growers

Certainly, growing hops for your own use doesn't demand a great number of plants. Most home growers

Hops want to climb. Installing a pole at the corner of the garden with a horizontal board at the top allows you to run several lengths of twine from the top of the pole to the ground. Once you train the bine to climb the strings the plant will take off. Illustration by Dahl Taylor

get by nicely with anywhere from one to five plants. But you still need to think carefully about where to establish those plants.

If you are thinking about planting hops in your garden, don't do it. Plant the hops somewhere else. When we first started growing hops as gardeners, it initially seemed to make sense to plant a hop in the herb garden outside the kitchen door. Hops are herb-like in that you can cook with them and use them for medicinal purposes. We envisioned the hop growing up the side of the house, hiding the electrical meter and the wire that ran to it. And grow up the side of the house it did. It also sent out a never-ending onslaught of roots, shoots, and runners under the ground that popped up everywhere and began to climb everything in an effort to colonize the entire herb garden, seizing territory even from the aggressive Kentucky mint and lemon balm that had long dominated the region.

It took us several years to successfully eradicate that hop plant. During this time we were forced to relocate the herbs, for their own protection, to a raised bed just inside the vegetable garden fence. Then, frustratingly, Dieter planted a salvaged hop rhizome in the new herb garden because he thought it would be nice to have the hop growing along the fence. Once again the battle ensued. Eventually we triumphed, this time not by killing the hop but by strictly training it to grow up a single tall pole that Dieter put in at the corner of the garden by aggressively and continuously cutting all shoots that come up anywhere else year in and year out. Over a period of about seventeen years the hop seems to have come to prefer the pole and, although it still sends out runners in the spring, it is

Small-Scale Hop Trellising

Over the years people have developed many different designs for hop yards, and of course small-scale plantings of five to ten bines for the home brewer or homesteader can be improvised by making use of existing structures and materials. The basic principles are the same whether growing hops on a small or commercial scale. The hops must grow upward, the higher the better, and have a support system strong enough to hold their weight at maturity. The material that the bines actually wrap themselves around should be made of something strong yet biodegradable that can be cut down at harvest time along with the bine. The bines must have a lot of space between them—at least 3 feet (0.9 meter)—so they don't become entangled.

Here are some simple trellising options for small-scale growers.

Hops grown on a small scale depicted here will provide a home brewer with an ample harvest.
Illustration by Dahl Taylor

- Create a trellis off the side of a building by running string or twine on a diagonal from the upper reaches of the building and anchoring it in the ground. For attaching to the building, tie the string or twine to an eyebolt screwed into the side of the building. Make sure this setup is strong enough to support about 100 pounds (45.4 kilograms). Even at full maturity the bine itself will not exceed a weight of 50 pounds (22.7 kilograms), but the extra weight capacity should provide for a rain-soaked hop bine being buffeted about in a storm. Anchor the line in the ground about 3 feet (0.9 meter) from the building using a stake.
- Erect a single pole, nail a horizontal board to the top of it to create a T shape, and (depending on the length of the horizontal board) run two or four strings from the top of the pole to the ground. The post should be at least 16 feet (4.9 meters) high, and to install it you'll have to dig a posthole 3 feet (0.9 meter) deep and bury the bottom of the pole in it to secure it in place. Again, anchor your strings to the ground with stakes to support the eventual weight of the hops.
- Erect a micro hop yard composed of a short row of four poles, anchored to the ground by cables, with a single wire running along the top to support several bines. Even though the hop trellis system is small in a micro yard, it should still be tall. Because the system will carry less weight, it does not have to be as strong as a full-scale hop yard trellis system. You want your poles to be at least 16 feet (4.9 meters) high, but you do not have to bury them as deep. Burying your poles at a depth of 3 feet (0.9 meter) will suffice. Put the poles 6 feet (1.8 meters) apart. Nail a 10-foot-long (3-meter-long) board to the tops of the poles, with supports coming off the top of each pole. Run cable from the top of the pole to the ground, where you can anchor it with a stake.

Whatever the small-scale design used, it is important to remember to leave as much space between hops at the top of the trellis as possible so that the bines do not become intertwined with each other. This makes it difficult to cut down the individual bines. It also creates a dense mass of vegetation that prevents the evaporation of moisture, creating a humid environment conducive to disease.

Picking hops by hand from a standing bine makes for a stronger hop plant. Because the cones are removed but the bine is left standing, the vegetation can continue to photosynthesize for the remainder of the growing season, generating energy that can be stored in the root system for the following spring.

and, with a proper trellis system, can be so productive you may not need many bines.

Because different styles of beer call for different varieties of hops, you will probably want to have several varieties to choose from; that means that even if you only plan to do a few batches, you will have to have more than one hop plant—but remember that the more bines you have, the more work you will have to do to take care of them. It is also important to take into account how much time you will have during the growing season to dedicate to the hops. Hop plants can be pretty high maintenance, especially in the spring and early summer. Poor care will reduce the yield. If you want to maximize the amount of hop cones you harvest, don't go crazy and plant more hops than you can handle.

Although it is fun to try lots of different hop varieties when home brewing, keep in mind that not all varieties grow well in all areas of the country. Growers in the eastern United States will want to stick with varieties that do well in a more humid climate. Although an obsession with many, home brewing is still supposed to be a hobby, and presumably you have a job. Try integrating hops into a brewer's garden that includes grains, herbs, and even small fruits such as berries that can all be used to brew beer made exclusively with homegrown ingredients.

Siting the Commercial-Scale Hop Yard

If you are going to try to make some money growing hops you are going to need a hop yard. A hop yard can be on a piece of land as small as ¼ acre (0.1 hectare) or it can be hundreds of acres in size. Whether large or small, a hop yard involves not just the plants in the ground but an elevated trellis system composed of poles and cable along with irrigation at ground level. With a hop yard as little as ½ acre (0.2 hectare) in size, you can produce a volume of hops sufficient to sell to home brewers or a small brew-pub. But to achieve a profit growing hops, you need to have a hop yard that is at least 1 acre (0.4 hectare) in size. A basic 1-acre hop yard contains approximately nine-hundred plants and when fully mature has the potential to produce 1,500 pounds (680 kilograms) of dried hop flowers, which currently sell for between $6 and $12 a pound depending on variety, quality, and market conditions.

The size of the hop yard will be determined by your long-term revenue and associated production goals. When calculating your production goals it is important to weigh your projected revenue against the costs of installing the trellis system, purchasing rhizomes, and acquiring the necessary equipment, such as hop-harvesting machinery and dryers. It will take five years for hop sales to repay the sizable investment involved in starting a hop yard.

How much you can afford to invest up front and how long you can wait to get paid back on your investment will be the most important factors when deciding what size hop yard you need to install in your first year. You may find it more feasible to work in stages, choosing your site to allow for expansion. Once you have determined the size hop yard you need, it is time to

The hop yard at Foothill Hops in Munnsville, New York. Foothill Hops, established in 2001, was one of the first farms to start growing hops again in New York State. Photograph by Laura Ten Eyck.

select the location. Because this location will be permanent, choosing the right site for your hop yard is one of the single biggest decisions you will make as a commercial hop grower. Whether you are planting one, one thousand, or ten thousand hop plants you must consider the same critical factors.

Sun

Thinking in terms of the big, big picture—as in where on planet Earth do hops grow best—it is between 35 and 55 degrees of latitude, which in the eastern part of the United States means from Maine to North Carolina. The reason has to do with both temperature and day length. Hops need a minimum of 120 frost-free days to flower. Go any farther north and it is too cold. Hops don't grow well near the equator because they don't like extreme heat and need a minimum day length of twelve hours (the length of daylight at the equator is twelve hours).

Hops do like sun and need lots of it. They also require significant changes in the length of periods of light and dark during the growing season. This response to light exposure is called photoperiodism—in other words, a reaction to the amount of time spent in light or darkness. It is the same phenomenon that makes some animals change color, migrate, and hibernate as periods of light and dark

No matter what size hop yard you are planning to install there are several critical factors, including sun exposure and drainage, that must be taken into account when choosing your location.

Wind can be both a friend and an enemy. Sited on a hill, this plant gets a good breeze, which can keep moisture out of the hop yard to help prevent disease and control insect outbreaks. But too much wind can put stress on the hop trellis system.

humidity. It's called evaporation—when the breeze blows, moisture dries up. People like it when evaporation happens because it is cooling. Hops like a breeze because they like their leaves to be dry.

That being said, too much wind is not a good thing. High wind tears up the leaves of the plants and can break the bines, slowing down growth and decreasing production. Also, as the plants grow they become quite heavy. A strong wind knocking them around, especially if they are weighed down with water, takes a toll on the trellis system. If the strings holding up the hop bines are not as taut as they should be, the bines will sway, straining the trellis even more. Take typical wind speed and direction into account when you are choosing your hop yard site and orienting your trellis system. Of course, if you live on or near the property where you are considering putting your hop yard you will already have an understanding of which way the wind blows and how fast. But it might be helpful to put up a windsock and monitor the situation. If you want to get scientific about it you can use a remote wireless anemometer to collect some data on various sites.

Planting your hop yard at high elevation increases access to breezes and also ensures good drainage. Another advantage of a higher elevation is that your hop yard will not be in a frost valley, where cold air that can freeze young shoots settles in the spring. Consider planting your hop yard on a gentle slope. Not only does this help with drainage but it also helps with air circulation. As the temperature rises during the day, the air heats and moves upward. As the temperature cools the air in the evening it moves back down the slope. Although this air motion is so slight it cannot be detected as a breeze it still counts as air circulation that helps the hop yard stay dry.

Accessibility

Accessibility is another important factor that should not be overlooked when selecting a site for your hop yard. Growing hops is labor intensive. In addition, a hop yard requires frequent monitoring for pests and disease. It is not practical for your hop yard to be in a remote location on your farm or to be a great distance from where you live. The site should also be easily accessible for equipment such as tractors and other farm machinery.

The hop yard should be conveniently located to the hop processing area. Once the bines are cut, the cones need to be picked right away. In some cases this can be done in the field, but it is more common to pick the cones from the bines in proximity to the dryer, usually housed in a barn or other farm structure. But wherever you harvest the hops, once picked from the bine they will need to be quickly conveyed to the drying area to ensure the best quality.

Introducing livestock into the hop yard after the harvest helps control weeds and adds manure to increase fertility.

If you plan to incorporate livestock into your hop-growing operation, accessibility is also a factor. Livestock such as sheep can be helpful in controlling weeds, as well as stripping lower leaves from bines to prevent disease moving up from the ground. If livestock such as sheep or chickens will be in the hop yard at certain times, you'll need to consider the ease of moving them in and out of the hop yard from the barn or pasture when choosing the site. If you plan to use livestock manure or compost for fertilizer, consider where the manure and compost is being collected and the ease of transporting it from there into the hop yard. One other issue to think about is where the hop yard will be located in relationship to the livestock's wintering area. Seeds blowing into the hop yard from hay feeding stations can really add to your weed problem. When we had alfalfa sprouting in our hop yard, at first we were confused, but then we realized that we had located our hop yard directly downwind of the barnyard where our livestock wintered.

Depending on your business model, marketing may also be a factor. If you are selling hops to home brewers, or using hops to make beer commercially, it is a good idea to have your hop yard in a location that is visible to the public. We have found that people are fascinated by hop yards. Keeping the hop yard highly visible is a great marketing tool and advertisement for your product.

Previous Uses

Consider previous activities that have occurred in your prospective location. It is quite difficult to take raw ground, such as pasture or orchard, and turn it into a hop yard. The weeds and grasses will be well established, and it will take a significant amount of time and energy to eradicate them. It is easier to establish a hop yard on a field that has previously been planted in corn or some other type of cash crop such as soybeans, where the weeds have already been brought under control. If you are considering growing hops organically you will need to know if nonorganic fertilizers, herbicides, and pesticides have been used on the ground. If so you will need to wait three years before you can apply for organic certification.

Soil

The quality of the soil is obviously a critical consideration when siting your hop yard, yet it must be weighed alongside all of the factors described above. Although it is clearly not a good choice to plant any crop in a completely inappropriate soil type, it is worth remembering that once a hop yard is installed it is highly impractical to move it if, for example, you find that even though a location's soil is great it is in a low area that collects water or sits smack dab in the middle of a high-velocity wind tunnel. While you can do little to change conditions such as elevation, water pooling, sun exposure, and wind direction it is within your power to improve the soil. For discussion on how to do that, and to manage the soil in a hop yard, see Chapter 4. In the meantime, keep these soil-related factors in mind when making site decisions:

- As mentioned earlier, it is critical that the soil for hops be well drained.
- It should also be deep, ideally 3 feet (0.9 meter) or more, to allow for the proper development of the hop's root system.

When it comes to the soil one of the most important factors is drainage. Since hops abhor standing water it is essential the soil be well drained. A healthy level of organic matter in the soil is also critical to enabling the hop plant to take up the large amount of nitrogen necessary to produce a viable crop of hop cones.

- Hops do best in a medium-textured soil with good tilth—such as sandy or silty loam.

When evaluating soils, start by checking out the soil maps for your county. The Natural Resources Conservation Service (NRCS), which is part of the United States Department of Agriculture, maintains soil maps for over 95 percent of counties in the country. You can look at the soil maps of potential hop yard sites online. For help in understanding the maps and data, determining your soil type, and conducting a soil test, reach out to your county's soil and water conservation district and land grant college Cooperative Extension offices.

If the soil in your location is not ideal, it can be improved over time through cover cropping and incorporating organic matter and nutrients in the form of compost and manure.

mature. A mature hop yard can consume between 100 and 240 pounds (45.4 to 109 kilograms) of nitrogen per acre during the growing season. In the spring, before plant growth takes off, nitrogen uptake is slow. During this time the plant is producing long shoots and very few leaves. This growth is fueled by energy stored in its roots at the end of the previous season. The plant will only use about 10 percent of its total nitrogen uptake by the end of May. The vast majority of the hops' nitrogen uptake occurs in the month of June. A fertility regime of 120 pounds (54.4 kilograms) of nitrogen applied in the spring with another 120 pounds applied through mid-July has been the industry standard in the Northwest. In general more nitrogen means a higher yield, but an excess of nitrogen can damage both the hops and the environment. An excess of nitrogen promotes excessively fleshy plant growth, which is vulnerable to disease. If you apply more nitrogen than the plant can take up from the soil, this nitrogen can wash away during heavy rains, contaminating aquifers and nearby surface water.

Learning about the Soil in Your Hop Yard

As noted in Chapter 3, when you are considering a potential site for your hop yard you will need to learn about the soils in that location, and a great way to get started is to consult the USDA soil survey for your county. Back in 1896 a massive effort was launched to survey soil types nationwide. Today the resulting USDA-led partnership and the database it manages is known as the National Cooperative Soil Survey. Soil surveys exist for nearly every county in every state in the country. A county soil survey is essentially a map of the county overlaid on an aerial photograph on which various soil types and slopes are indicated. It includes a description of the soil type as well as an interpretation of how appropriate the soil is for a variety of uses—from various crops to wildlife habitat or building construction. The soil survey is available online (see Resource section), and in many cases county soil surveys are still available in book form and can be obtained from your county's soil and water conservation district office.

Whether you are looking at the soil survey maps online or on paper, the first step will be to find your piece of land. On the map the soil in your planned location will be labeled. For example, on the soil survey for our hop yard location the soil area is labeled *Chc*. The *Ch* stands for Chenango, which is the soil description. The lowercase *c* indicates the degree of slope. Once you know this you can then look it up in the book and read a rather extensive description of the nature of this soil. You can also consult a chart and see what your expected yield in bushels can be for select commodity crops as well as other information.

Once you have consulted the soil survey, the next step is to do an actual soil test to establish the various levels of nutrients in the soil. Testing your soil involves taking soil samples from various representative sections of the site and mixing them together to create a homogenous sample. You then send this sample along with a form on which you supply information about the soil, the history of the site, and what you plan to grow to a laboratory. At the laboratory the soil will be tested, and you will find out about the pH of the soil as well as the levels of organic matter, various nutrients in your soil, and what you need to do to adjust them to appropriate levels for the crop you intend to plant. After the initial soil test, the general recommendation is that you test the soil every three years, but if you are working to make major changes you may want to conduct a test each year. Bear in mind, though, that it is important to test for pH at the same time each year for consistency.

Filling out the form that will accompany the soil to the laboratory correctly is just as important as collecting the sample correctly. The form will ask you to provide identification for the field from which the soil sample came. If you only have one hop yard this may seem silly, but it is good protocol to name the locations of your plantings. For example, on Indian Ladder Farms, different sections of the apple

The first step in learning how to care for your soil is to have your soil tested to find out what it needs. When you submit your soil sample make sure you indicate the type of soil as well as the crop you plan to grow to help the laboratory tailor their recommendations to your needs.

The soil's pH can be changed by adding materials to the soil. For example, if the soil is too acidic you add lime. If the soil is too base you add sulfur. Soil chemistry can get pretty complicated. Fortunately, when the results of the soil test come back to you, the lab not only tells you what your soil pH is, it tells you what you need to do to the soil to get it to the correct pH. Be sure to ask your Extension agent to help you interpret the recommendations from the lab and decide what to apply to your soil, as there may be extenuating circumstances. For example, different types of lime contain varying levels of micronutrients, depending on where the lime is sourced. If you have an issue with micronutrients and you make the wrong choice, you could exacerbate your micronutrient problem.

In addition to helping you get the pH where it needs to be, the soil test also tells you about the levels of organic matter and nutrients in the soil and provides information about what to apply to meet

Take advantage of the time before your hop rhizomes go in the ground to make field-wide soil adjustments, such as application of lime to adjust pH and the addition of organic matter through cover cropping.

Cover crops, such as the red clover pictured here, are a great method for suppressing weeds on the site of the future hop yard prior to putting the hops in the ground. When turned into the soil the cover crop is also a source of nitrogen and organic matter.

the needs of your crop. While you are working to adjust your soil's pH you can address other issues identified by the soil testing. If your soil is lacking in organic matter it makes sense to work in composted livestock manure, then plant a cover crop. The cover crop can also block weeds and nourish the soil.

Cover Cropping

Whether you are planting a small number of hop plants behind your house or installing a 1-acre (0.4 hectare) hop yard, consider cover crops before you put your rhizomes in the ground. There is nothing more frustrating than working hard to install your hop yard trellis system, plant your rhizomes, prune shoots, and begin training your bines only to have them be overwhelmed by the alarming growth rate of the previously established weed population.

Cover crops in general serve a variety of purposes. In their simplest form, they are planted to protect a field from wind and water erosion during the dormant season after the primary crop has been harvested. They also help the soil retain moisture. When a cover crop is tilled under, it adds fresh organic matter into the soil, acting as green manure. This will improve the soil's tilth, increase the level of biological activity in the soil, introduce nutrients, and fix nitrogen. Also the very act of growing a cover crop can help improve compacted soil as the cover crop's roots penetrate and aerate the soil.

Cover crops can also act as "smother crops," planted for the primary purpose of controlling weeds. Close and fast-growing cover crops, such as buckwheat, will literally smother weeds by taking over every square inch of available space and choking them out. Some types of cover crops, such as rye, also have allelopathic properties, meaning they produce a toxic chemical that inhibits the growth of weeds. Cover crops can also be planted to interrupt the life cycles of insect and disease pests damaging a commercial crop as well as to provide habitat for pollinators and beneficial insects.

Many commercial crops such as corn are routinely rotated with cover crops to accomplish some or all of the above goals. Since hops are a perennial crop it is not possible to "rotate" them with other crops in the traditional sense. That does not mean that cover cropping is not an important tool for hop growers, but the main role played by the cover crop is in setting the stage for a healthy hop yard.

Develop your cover crop plan based on the quality of the soil and the conditions in the area in which you will plant. Maybe the field you want to grow on has produced crop after crop of feed corn and is now low in organic matter and depleted of nutrients. Maybe it is a former cow pasture that is being turned over for the first time. Each of these scenarios and the many other possibilities come with their own special needs. Not only will you need to choose a cover crop that will accomplish your goals, you may need to choose a series of cover crops that will be rotated seasonally over a period of a year or more prior to putting your hop rhizomes in the ground.

Even though the cover crop is not the end goal but a step toward that goal, it is still a crop and needs care to thrive and serve its purpose. Once you choose the type of cover crop you need, take the time to find

Suitable Cover Crops to Grow the Year before Planting Hops

Winter Rye	Ryegrass	Buckwheat
Oats	Barley	Marigold
Wheat	Sweet Clover	Sudan Grass
Vetch	Red Clover	

Source: *2014 Cornell Integrated Hops Production Guide*

These buds, fed by nutrient-rich soil, are emerging from a hop rhizome. Fast-growing hops take up a large amount of nitrogen from the soil. Nitrogen and organic matter can be added to the soil in the form of compost and livestock manure.

The amount of organic matter in your soil also contributes to how much nitrogen you need to apply. A high percentage of organic matter in the soil not only contributes to the nitrogen available to the plant, it also makes the environment more conducive to the effective delivery of nitrogen to the plant. You really can't go wrong by building the organic matter in the soil in your hop yard through the application of manure and compost. But if you plan to rely on manure and compost for the bulk of the nitrogen you are providing to the hops during that critical growth stage in May and June you are going to have to do some planning.

Manure and Compost

You can apply raw manure, composted manure, or plant-based compost to your hop yard. If you are using raw manure you have to stop applying it a minimum of 120 days before you begin to harvest cones. As a general practice, side-dressing the hop plants with manure, working it into the soil before you plant, or tilling it into the soil in the hop yard's rows improves both the level of organic matter in your soil and the hops' nitrogen uptake.

To apply fertilizer to hops through side-dressing, lay down a layer of compost or raw manure several inches thick on the surface of the ground extending about 2 feet (0.6 meter) in diameter from the base of each plant. You do not need to work the side dressing into the soil. As a matter of fact you *should not* do that because digging in the ground around the base of the plant will disturb the crown. The nitrogen in the compost or manure will work its way down into the soil and provide an extra boost to the plant during the

These hop shoots are emerging from plastic mulch. Nitrogen sourced from manure tea and fish emulsion is delivered via irrigation hoses running beneath the plastic.

growing season. It is best to side dress hops with fertilizer in late May to give them the energy they need for the big growth spurt to the top of the trellis that takes place during the month of June.

One handy way to replenish nitrogen and other nutrients in the soil while building organic matter is to compost the hop bines themselves once the cones have been removed, then add that compost back into the soil. Many commercial hop yards do this. It is important that the composting be done correctly. You'll want to achieve a high enough temperature to kill any destructive insects and disease pathogens that may exist on the cut bines so you do not reintroduce these into the hop yard. To kill bacteria the Rodale Institute recommends compost be maintained at a temperature of between 131 and 170 degrees Fahrenheit (55 and 76.7 degrees Celsius) while being turned five times over a period of fifteen days.

If you are fertigating—adding to your soil via your irrigation system—and you want to stick with purely organic materials, you will likely go with liquid forms such as manure tea or fish emulsion.

If you are planning to use manure or compost as your principal source of nitrogen during the hops' June growth spurt you need to have it tested so that you can find out how much nitrogen it contains. Manure test kits for raw, composted, and even liquid manure (think dairy farm manure lagoons) are available from your county's Cooperative Extension office. The same labs that do soil testing generally do manure and compost testing as well.

This test is important because the levels of nitrogen in composted manure differ from that in raw

Hops grow the most between mid-June and mid-July. Side-dressing the hop plants with compost or manure in early June helps ensure the plants have access to the nitrogen they need to fuel this growth.

Plant Tissue Testing

Once your yard is established and you begin growing hops, you can also test your plants for clues on how to best manage the nutrients in your soil. Plant tissue testing is a way to find out what types and quantities of nutrients the hops are actually drawing out of the soil during the growing season. This is called "uptake data." Plant tissue testing will tell you whether your fertilizer plan is working. It can also help you diagnose nutrient deficiencies that may be negatively impacting the hops. The lab analyzing the plant tissue can test for an array of nutrients or just for nitrogen, depending on your needs. No university-developed standards for nutrition levels in hops tissue exist, but tissue analysis can be helpful for measuring the results of fertilizer applications comparing the current year to the previous year.

In general when plant tissue is tested for nutrient content, the most recently fully matured leaves are collected. In hops in particular, the testing is done on the part of the plant called the petiole—the leaf-stalk connecting the leaf to the stem. To get useful results, it is important to do accurate sampling. This means collecting the right part of the plant at the right time.

To assess the amount of nutrients the hop has taken up from the soil, Agro-One Soils Laboratory calls for fifty to sixty petioles to be collected from hop bines throughout the hop yard at a height of between 5 and 6 feet (1.5 and 1.8 meters). If possible, collect petioles from leaves that have recently matured as

Sending plant tissue to a laboratory for testing will help you know how much nitrogen your hops are actually extracting from the soil. The part of the hop that is tested is called the petiole, the point where the leaf connects to its stem.

opposed to those that are immature or old. Avoid leaves that are damaged by insects, disease, cold, heat, or moisture stress. This will measure how much nitrogen your healthy plants are consuming. If you are trying to diagnose a possible nutrient imbalance in the hop yard, take samples from both healthy and afflicted bines. Comparing the two will help you learn if the afflicted plant's nitrogen level might be contributing to its problem.

Your sample can be sent to the lab fresh or dried. Whichever lab you choose, make sure they send you paper plant-tissue sample bags before you collect your samples. Don't use paper lunch bags available from the store, as these bags are treated with the fire-retardant borax. The presence of borax can throw off the boron test results. The fresh sample should weigh about 2 ounces (56.7 grams) and be packaged in a box with an ice pack. If dried, the sample should constitute about 2 cups (473 milliliters) lightly packed. Dry samples do not need to be kept cold. Write an identification number on the sample bag, and use that same number on the form you'll need to send back to the lab along with the sample. The form will ask you to choose between a test for a broad array of nutrients or a test for just nitrogen.

If you are simply trying to understand the level at which the plant is actually consuming the nitrogen you are providing, you can stick with the nitrogen test. If you want to know the levels of a spectrum of nutrients, go for the array. In some cases plants that look like they might be diseased

Installing a hop yard is hard work. To accomplish the task of erecting the poles we relied on a posthole digger, a tractor, and a lot of friends.

CHAPTER 5

Small-Scale Hop Yard Design and Trellising

The hop yard closest to our house consists of four rows of 18-foot-high (5.5-meter-high) locust wood poles planted 4 feet (1.2 meters) in the ground and interconnected by a grid of strong cable. Erecting it was no small feat. And often when I walk through it, I think back to when its infrastructure was installed using a posthole digger, a tractor, and the brute strength of Dieter and several friends.

It isn't hard to attract our volunteer work force. For one thing they are all good friends and like to do things together. For another thing, although most do not have farms of their own they are all interested in

Sinking poles 4 feet (1.2 meters) into the ground—so they'll hold up to wind and the weight of the hops—requires teamwork. Photograph by Laura Ten Eyck

Trisha Putman, an avid organic gardener, has helped out in our hop yard in many stages—from installation to harvesting.

food and farming. They have their own big gardens, or work with us in ours. We collaborate with their families to raise lambs and broiling chickens here on our farm, sharing costs, labor, and meat. They are also committed customers of our son's egg business, coming and going from our barn regularly to load up on dozens of eggs, filling the lunch box on top of the refrigerator with cash. But there is one technique above all others that always produces a volunteer workforce, and that is a cold keg of beer in the barn refrigerator. Once the work is done, and often before the work is done, the beer begins to flow. What we may lose in speed and productivity we make up for in quality of life.

Even with friends who are willing to rally, a hop yard is labor intensive. And perhaps the most labor intensive time is at the very beginning, when you install the system that will support your hops for years to come.

Hop yard infrastructure has to be built big and strong to support hundreds of mature hop plants that cumulatively weigh tons.

Designing a Commercial Hop Yard

While you are working on adjusting your pH, improving your soil, and planting your cover crops, you can be thinking about what kind of trellis system you want to install. The modern hop trellis is a network of wooden poles, cable, and hardware that create a vertical infrastructure to support the hop bines to enable them to reach maximum production. The trellis system infrastructure must be extremely strong and secure because the hop is such a large, heavy plant. A mature plant at harvest can weigh as much as 50 pounds (22.7 kilograms). When you are talking about 1 acre containing nearly 1,000 plants, this adds up. For example, each row of our 1-acre (0.4 hectare) hop yard has 215 plants. With four rows, that's 860 hop bines; if you do the math you realize that the total plant load has the potential to reach a weight of over 20 tons (18.1 metric tons). That's really heavy, especially when you are talking about a mass of vegetation that can not only become waterlogged but move when the wind blows during a rainstorm.

Back in the olden days people did not use cables for trellising. Instead they planted tall wooden poles in the ground and trained the hops to grow up them. When it came time to harvest, a man called the "tender" would pull the pole out of the ground and lay it horizontally on a rack suspended over a bin. The pickers would pluck all the flowers off the bine and drop them into the bin below. The bine, stripped of its flowers, would be pulled from the pole and the pole would be set aside until the next growing season, when it would be used again. The problem with this system of using one pole per bine was that it added up to a lot of poles, especially as hop yards got bigger and bigger. Even though the poles were reused they only lasted about three years; then more poles had to be found. As all the trees were cut down in the area hop growers had to import poles from farther and farther away. This became really expensive so people started experimenting with using fewer poles, stringing a wire between them and growing the hop up strings that hung down from the wire.

Hop Yard Layout

A conventional hop yard trellis system is essentially made up of a checkerboard pattern of poles interconnected with cables running from pole to pole along each row, then crosswise horizontally for added support. Strings hang down from the trellis cable to the ground. The hop bines climb the strings. Poles are spaced between 20 and 50 feet (6.1 and 15.2 meters) apart. Hops are planted 3 feet (0.9 meter) apart. For example, if your poles are 30 feet (9.1 meters) apart you will have ten strings hanging at a distance of 3 feet apart between each pole. Some growers opt to add floating rows to the trellis system to increase the number of plants it can support (as described under High-Density Hop Yards below).

There are a lot of variables to consider when laying out the hop yard. Some of these choices, such as the length and width of your hop yard, will be dictated by the dimensions of the location you have chosen. Other decisions will be based on the level of strength you want in your trellis system, which will be determined by the size of the mature bines typical of the varieties you have chosen to plant, whether you are going organic or conventional, and weather factors related to your site—such as the potential for high wind and heavy rain.

One of the first decisions you will make is how high your poles will be and how deep to bury them. As mentioned earlier, 18 feet (5.5 meters) is the recommended height for a trellis system for hops to reach maximum productivity. That being said, if you are planning on an organic hop yard, some growers have recommended to us that you can get away with shorter poles, as hops not fed synthetic nitrogen simply will not grow to be as big as those that are; so for an organic hop yard 16 feet (4.9 meters) in height may suffice. The other consideration is the depth at which the pole will be buried. We have seen people burying poles at depths between 2 and 4 feet (0.6 and 1.2 meters). Although burying poles at a depth of 4 feet is strongly recommended, in some cases people have gone with a shallower depth because the ground in the hop yard was so rocky they could not dig any deeper. More depth means more strength in the hop yard, and strength is critical.

Once you've settled on the length of your poles it is time to consider the spacing in between the poles. You must decide the distance between each pole in a given row as well as the spacing between rows. Poles are expensive and hard to put in so you don't want to use more than you have to, but on the flip side if you don't install enough poles there won't be sufficient support for the trellis holding up the bines. In our research we came across the previous recommendation for spacing between poles in the rows ranging from 20 to 50 feet (6.1 and 15.2 meters).

There are several things to consider when deciding how far apart your poles should be. One consideration is the size of the pole. If you are using smaller poles, what they may lack in strength can be made up for in numbers. Also, if you are unable to bury your poles to a full depth of 4 feet, more poles can help carry the weight of the bine. Another thing to consider is how long your rows are. Very long rows will require extra strength, which can be achieved by a higher density of poles.

The next consideration is how wide you want your rows to be. The standard row width in a hop yard is 14 feet (4.3 meters) on center. This means that the distance between the center of one planted strip of hop crowns is 14 feet from the center of the adjacent planted strip. The hops themselves take up a strip

End poles, also called anchor poles, are critical load bearers for the hop yard trellis system. We used discarded utility poles for our end poles and black locust trees, seen piled here, for the poles in the rows.

about 2 feet (0.6 meter) wide. Using the standard width provides open lanes that are 10 feet (3 meters) wide, large enough for equipment such as tractors to move up and down the rows easily.

The hop yard's outside poles, which are there to strengthen the trellis system, can be the same height as the row poles. Many choose to angle the outside poles away from the hop yard to increase their ability to support the trellis. The outside poles are then anchored to the ground with guy wires connected to ground anchors. There are many different types of ground anchors to choose from. Some are embedded directly in the soil and others in concrete or stone.

High-Density Hop Yards

Growers that want to increase the density of a hop yard can add floating rows between the poles. A floating row's cable runs crosswise from wire to wire at a midpoint between two poles. These additional wires allow a grower to train two bines off a single plant, doubling the production of each crown. The strings running from these floating-row cables angle and anchor in the ground beside the hop crown, forming a V. The grower then trains two bines from one hop plant, each climbing one of the strings forming the V.

Increased density can also be achieved by adding more plants to a given space. This can mean planting hops 2 feet (0.6 meter) apart or less within the rows or sometimes even planting additional plants in the aisles that run between the rows. The choice to increase density by planting more hops in the same amount of space must be made carefully. Whether or not it will work depends on the variety of hop. Some hop varieties have very long sidearms,

This newly erected hop yard, Hager Hops, located in Cooperstown, New York, features outside poles angled away from the hop yard to increase strength as well as a trellis system with floating rows.

and if they are planted too close together they will easily become entangled. Also, if hop plantings become too dense the hops may not get enough sunlight; from a combined lack of sun and air circulation, moisture levels can rise, creating a humid environment conducive to the spread of disease.

Low Trellis Design

To reduce costs, a number of growers are experimenting with low trellis design. People often consider the word low to mean something close to the ground or at least shorter than a person. But in the high-elevation world of hops the word "low" refers to a height of 9 feet (2.7 meters). A low trellis is essentially half the height of a high trellis. The same amount of cable is necessary, but the poles are obviously smaller and therefore less expensive. The reduced investment makes it easier for the commercial hop grower to put up a hop yard to grow a new variety when the market demands.

The problem with this thinking is that right now the market is demanding extremely high-alpha hops, which are very large, fast-growing plants that don't do well on a low trellis system. Some traditional varieties with high alpha levels such as Cascade, Centennial, and Galena can be grown on a low trellis, but it takes a bit of trickery on the part of the grower. The top part of the plant where growth is occurring is called the apical meristem. It is here that the plant manufactures hormones that drive growth. When the apical meristem hits the 9-foot top of the low trellis it wants to grow still higher. At this point the apical meristem must be snipped. This cuts off the flow of hormones, halting the climb upward and stimulating the growth of the sidearms on which th flowers will form. That is extra labor—and even though it takes place at a height less than 18 feet (5.5 meters), 9 feet still requires you to get up in the air. Even if you take this step, most hop varieties on a low trellis still do not produce as many cones as they do on a high trellis. Older varieties can produce well at lower levels, but new varieties tend to be more productive at the upper end of the range.

This problem can be avoided by planting true dwarf varieties of hops, which are being bred by grower groups such as the American Dwarf Hop Association, as well as the United States Department of Agriculture. True dwarf varieties include First Gold, Pioneer, Herald, Sovereign, Boadicea, and Pilot.

Instead of climbing a string, as with high trellis hops, low trellis hops are generally grown on mesh panels that run from pole to pole. This system requires a different harvesting method. In a high trellis hop yard a mechanical top cutter, a machine that cuts the string on which the bine is growing from the top of the trellis, travels up and down the ows cutting the bines and dropping them into a uck for transporting to the harvesting machine. p bines on a low trellis system are grown on mesh harvested in the field by a machine that sandes the bines between two panels, which pluck ones and leave the bines standing. This method vesting is actually better for the plant—instead g cut down, it gets to continue to photosyn- and build up nutrients for the winter. ized harvest of a low trellis hop yard requires ed equipment, but for the small-scale ho intends to hand harvest, the low trellis ld be perfect.

ng

e decided on your design and sketched is important to choose what varieties nd where. It is important to keep each to its own area to reduce potential vest time. It makes sense to lay the varieties out in the order that they will be harvested, planting early maturing bines in the first row and later maturing bines in the last row so that you are not "hopping" around the hop yard at harvest time, creating confusion. In our 1-acre (0.4 hectare) hop yard we are growing Brewer's Gold, Cascade, Centennial, Cluster, and Nugget. In an ideal world, each row would be composed of only one variety for ease of harvesting. We were able to adhere to this in the beginning, but in our second year we ended up planting two varieties in one row.

Our hop yard design is based on the conventional trellis design method. Our four rows are each 650 feet (198.1 meters) long, and we chose to install one pole every 35 feet (10.7 meters). We worked with 22-foot-long (6.7 meter) poles, burying them 4 feet (1.2 meters) deep in the ground for a standing height of 18 feet (5.5 meters). We went with the standard row width of 14 feet (4.3 meters) on center. Because they happened to be available to us, we ended up using discarded utility poles as our end poles. There are concerns for organic growers about using utility poles in the hop yard as they have been treated with preservatives that may be toxic. Although we are not certified organic this was a concern of ours. The utility poles are only used as anchor poles, do not come into contact with the hops themselves, and are approximately 15 feet (4.6 meters) from the soil under cultivation. Only a few feet farther away are the much newer utility poles alongside the road. Our discarded utility poles are much larger than our row poles: 24 feet long (7.3 meters) and 14 inches (35.6 centimeters) in diameter. We have them buried 6 feet (1.8 meters) in the ground, leaving 18 feet above ground, the same height as the interior poles. These end poles are anchored with a ground anchor buried 4 feet deep and connected with a 5⁄16-inch 1 × 7 type cable (to learn more about cable, see the Cable section under Materials below). To string the trellis we used cable of the same type and size. A single strand of cable runs along the tops of all the row posts. Currently we are only growing one bine from each

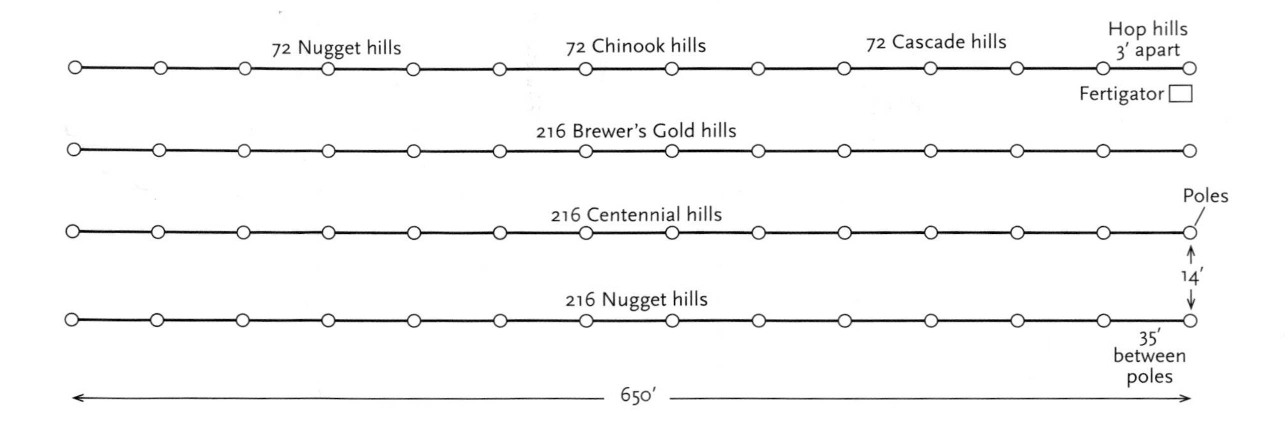

When planning a hop yard it is important to get your design on paper to help guide construction. Our design sketch shows the row layout, pole placement, and where different hop varieties are planted.

hop plant and only have one line of cable running between poles. We have not yet installed the trellising necessary for floating rows. When our bines are fully mature we will install the angled poles along the sides of the yard, string horizontal crosswise running cable, and install floating rows.

Materials

In their ongoing effort to maximize yield per acre growers have adapted a number of variations in layout and design, but all commercial hop yards make use of the same basic set of materials: poles, cable, ground anchors, and hardware.

POLES

Hop poles form the solid infrastructure of the hop yard, and depending on the size of your hop yard, you are going to need quite a few. Choosing and obtaining hop poles is a big deal. Ordering poles from a lumberyard is expensive. Finding cheaper, local alternatives can be difficult and time-consuming. This is something you can be researching while you are waiting for your cover crop to grow. You will need two types of poles. Anchor poles, also called end poles or outside poles, are the poles installed at the end of each row. The sole function of the anchor poles is to provide strength. Because of this, anchor poles can be bigger around and taller than the hop poles in the interior of the hop yard, referred to as inside poles.

Although some people putting up smaller hop yards have experimented with using row poles made of metal or PVC piping the convention is to use poles made of hardwood. There are several factors to consider when choosing poles, including height, diameter, straightness, and wood type.

Height. Today's industry standard is 18 feet (5.5 meters) because research has shown that the highest yields of hop cones are achieved when the hop bine is allowed to reach a height of 16 to 20 feet (4.9 to 6.1 meters), depending on the variety. Since each pole should be embedded 4 feet (1.2 meters) deep in the ground, you'll need to procure inside poles between 20 and 24 feet (7.3 meters) long, depending on how high you want your hops to get. The inside poles in our hop yard are 22 feet (6.7 meters) long, with 4 feet below ground and 18 feet above ground.

Diameter. To be strong the poles also have to be fairly thick. However, since they are already so long, you don't want them to be too thick or else they will be incredibly heavy. End poles should be 6 or 7 inches (15.2 to 17.8 centimeters) in diameter, while the inside poles should be between 3 and 5 inches (7.6 to 12.7 centimeters) in diameter, depending on their height.

Straightness. It is important that the trees you select for the hop yard poles be as straight as possible. Crooked poles will diminish the stability and strength of the trellis system.

WOOD TYPE

Poles can be round, essentially fashioned from tree trunks, or they can be made from split wood. The main thing is they need to be of a type of wood that is going to last a long time. Recommended types of wood include cedar, black locust, tamarack or other larch trees, and oak. It is important when considering the wood for your poles that you try to get something long-lasting to minimize the frequency with which you will have to replace poles in your hop yard. Black locust (though heavy) is thought to be the longest lasting, followed by cedar, then larch. Some people strip the bark from the wood and have it dipped in preservative to extend the poles' life. It is not necessary to use wood preservative if you use a long-lasting wood. It is important to avoid the use of wood preservative, as it is toxic and leaches into the soil. If you are growing hops organically this will not be an option.

Depending on where you get them, hop poles can cost between twenty and fifty dollars each—not

Pile of black locust poles ready to be installed in our 1-acre (0.4 hectare) hop yard. We purchased eighty poles at a cost of twenty dollars each.

including the cost of transporting them to your hop yard. It is best to get them locally. They are large and heavy, and you will need a lot of them, so transportation costs will increase the farther they have to travel.

We were fortunate to be able to obtain black locust locally for the poles in both of our hop yards. When we put in our pilot yard, we were able to connect with a woodsman not far from us. He had been referred to us by Cornell Cooperative Extension, and was willing to cut young black locust trees he had growing in a wood lot. Dieter went over and helped him cut down the trees and load them onto a wagon. We purchased nine poles at ten dollars each. We picked them up ourselves so we did not have to pay for delivery.

When seeking poles for our 1-acre hop yard we had similar good luck. The man we were hiring to build our brewery referred us to a local logger who was clearing woods to expand a cemetery in the next town over. It turned out there was a lot of black locust in the woods of exactly the right height and diameter—and it was straight, too. Dieter went over and tagged the eighty trees he wanted to have cut for poles, and they were delivered in three separate loads right to our hop yard. The poles cost $20 each for a total of $1,600, and we paid $200 for delivery.

We were happy to source our poles this way because in addition to being affordable it was also relatively sustainable. Cutting down woodlands that contribute to open space, outdoor recreation, and wildlife habitat is not something we necessarily want to be a part of. However, the land our eighty-pole purchase came from was owned by and immediately adjacent to an existing cemetery. An

A stand of black locust behind a nearby cemetery. A local logger who was removing the trees for an expansion of the cemetery allowed us to tag the trees we wanted to use as hop poles, then cut them for us.

expanded cemetery was preferable to a housing development, shopping mall, or parking lot. And as the older parts of the cemetery demonstrated, trees, nature, and wildlife reclaim the space, and the winding paths provide a peaceful place for people to walk. It turned out my mother had purchased her own gravesite in close proximity to the expanding edge of the cemetery, so one day we will find ourselves walking those paths. In addition black locust itself is considered a very sustainable source of wood because it grows quickly and regenerates itself. It is actually promoted by the Rainforest Alliance as an alternative to tropical woods for furniture. The tree sends up new shoots from its roots. When the parent tree is cut down, the roots send up even more shoots. If you have the space and are thinking about growing hops over the long term, you may want to consider integrating black locust into a farm woodlot to create a source for replacement poles.

Removing bark from black locust poles is here done with water blasted through a pressure washer. It is important to strip bark from the section of the poles that will be underground to stave off rot.

As with everything to do with growing hops, when purchasing poles, the further you plan ahead the better. If you are buying your poles straight from a woodlot it is best to have them cut and delivered a year before you are going to use them. This allows the wood to dry. Dry wood means lighter poles. In addition, when the wood is dry the bark loosens and falls off, or can be easily removed—important whether you plan to treat the wood with a preservative or not, since the lower part of even an untreated pole should be stripped of bark before it is buried. If it isn't, it will rot faster.

Because we did not plan as far ahead as we should have, our hop poles were green when we put them in, and we had a lot of difficulty when we tried to strip the bark before putting the poles in the ground. The first mode of attack was a machete. Not surprisingly this proved too difficult. We then tried an antique bark spud borrowed from a friend. This also proved difficult and time-consuming. Dieter and another friend, a professional painter named Bill, then tried to remove the bark by blasting the poles with water from Bill's pressure washer. This did work but took a very long time, and there was an unanticipated negative outcome. Black locust lasts so long because its wood and bark contain natural toxins that protect it from rot. When pressure washing the black locust bark from the poles, Dieter ended up soaking himself with water and, as it turns out, black locust toxin, which aggravated his eyes and gave him an itchy rash much like poison ivy.

CABLE

You will need cable to anchor your outside poles to the ground and also to run between the tops of the inside poles to form the elevated trellis that will support the hops' vertical growth. The strength of the cable you are going to need will depend on the length of your rows and the number of plants in each row.

Cable is used to anchor end poles to the ground. It also runs along the top of the poles, forming the elevated trellis that will support the hops. Photograph by Laura Ten Eyck

Keep in mind that cable is different from wire, even though the words are often used interchangeably. Wire refers simply to a single strand of wire. Cable is made of multiple strands of wire, just like a rope is made of strands of fiber. In fact, cable is sometimes referred to as "wire rope."

A perfect example of the confusion between wire and cable is "guy wire," also called a "guy." This "guy" is actually a cable, and because it anchors your end poles, which support the whole structure, it should be strong. In general it is recommended that aircraft cable be used in hop yards. The first thing I think of when I hear about aircraft cable is what military jets hook onto when landing on aircraft carriers. Aircraft cable is not as strong as that. However, aircraft cable is a special strength cable designed for the aircraft industry as well as military use. Because of its strength aircraft cable has been adopted for a variety of commercial uses that involve lifting and moving heavy things. It comes in a range of thicknesses and constructions and can be made of different metals. Steel is the preferred metal for hop yard trellising, and you can choose between galvanized and stainless steel depending on conditions. Galvanized steel is coated with zinc to protect it from corrosion and is approved by the Food and Drug Administration to be used in contact with food—except for acidic fruits such as citrus, which erode the zinc coating. Factors to consider when purchasing are the weight of the cable itself, its working load limit, breaking strength, flexibility, and fatigue resistance.

The strength of cable basically depends on the number and strength of the wires it is made of, which is referred to as its construction or type. For example, a cable made of seven individual wires wrapped together is referred to as type 1 × 7. A stronger type of cable made of seven different cables, with each cable containing seven individual wires, is called 7 × 7. An even stronger type of cable made of seven cables composed of nineteen wires each is called 7 × 19.

Cables also come in different diameters, which are measured in inches or their fractions. For example, a 7 × 19 type cable can come in sizes anywhere from 3/32 of an inch (.24 centimeter) to 3/8 of an inch (.95 centimeter) in diameter. The more cables and wires within a type of cable, the bigger it is in diameter and in general the stronger it is. Another dimension to consider is that cable made from more numerous, thinner wires is more flexible than cable made from fewer, thicker wires. Confusing? Yes. There are people whose whole careers are devoted to understanding and working with cable, and if you can find one or two of them to help you trellis your hop yard you should. Otherwise, you need to do some calculating.

To figure out what strength cable you will need, you need to know how long your rows are going to be and how many plants you are going to have in each row, factoring in the weight of each of those plants at maturity along with the weight of the cable itself. Each type and diameter of cable has a different breaking strength, which indicates the weight it can carry before snapping. Hop plants are generally planted 3 to 4 feet (0.9 to 1.2 meters) apart. So if you have a row that is 300 feet (91.4 meters) long and contains eighty five plants you are going to need a cable that can bear 3,400 pounds (1,542 kilograms) in plant weight, meaning you are going to want a 7 × 19 type cable at least 3/16 inches in diameter, which is really the minimum strength that should be used in a hop yard. You will use a heavier strength construction for the guy wire cable that anchors the end poles to the ground than you will use for the cable that forms the trellis system running along the tops of the poles.

Cable is priced by the foot, and the cost adds up fast. Depending on its type and diameter, cable can cost between 16 and 40 cents a foot. For a 1-acre hop yard this can add up to two thousand dollars or more. But there is no point in skimping on strength, as you will just make more work for yourself when your hop yard collapses. In fact, some of the growers we met with in the Northwest recommended eastern hop growers make their trellis systems significantly stronger than those out west because of all the rain and wind we get. Better safe than sorry. When a hop yard collapses, it is a real disaster.

Charts are available to help you to determine how much weight different types and sizes of cable can bear (an example from Schmidt Farm in Farmington, New York, is listed in the Resources).

CABLING HARDWARE

Hop yard hardware includes all the metal gadgets that attach the cable to the poles and keep it taut. Many farmers erecting a hop yard may have had experience running fence line or low-height trellis systems for fruit trees or grapes. Few have experience with high elevation trellising, which is really more akin to running utility lines than an agricultural activity. We have seen several different techniques recommended, and since most farmers are mechanically inclined and like to figure out their own way of doing things, we have found even more variety in the field. We were fortunate enough to have a couple of neighbors who have worked together for many years running utility lines and when it came time to put up our trellis system we turned to them for guidance. Below is a list of the hardware we used in our hop yard upon their recommendation. It may vary somewhat from what others have done, but the basic principles remain the same. How many of each piece of hardware you need depends on the design of your hop yard and the number of rows and poles. As with cable, the hardware comes in different materials and sizes. You have to calculate what you need based on the anticipated weight the hardware will have to bear. The cost of the hardware is going to depend both on the size of the hardware and what it is made of. Obviously the stronger the hardware, the more it is going to cost.

Ground Anchors. The biggest and perhaps the most critical piece of hardware is the ground anchor that secures the guy wire supporting the end pole to the ground. Various types of ground anchors, also called earth anchors, are available. We used auger-style ground anchors. This style anchor is essentially a giant metal screw that drills deep into the ground and is held in place by the soil around it. It is installed in the ground with an anchor adapter attachment on a mechanical earth drill. This type of ground anchor is not recommended for use in sandy soils, where it can loosen from the ground. Some reinforce the ground anchor by embedding it in concrete rather than soil.

Guy Strand Tensioner. The guy strand tensioner is the piece of hardware that connects the base of the guy wire to the ground anchor. It is made up of a long U-shaped piece of metal that hooks through a hole at the top of the ground anchor and is fitted with a

The ground anchor is a screw-shaped length of metal that drills into the ground, anchoring the guy wire.

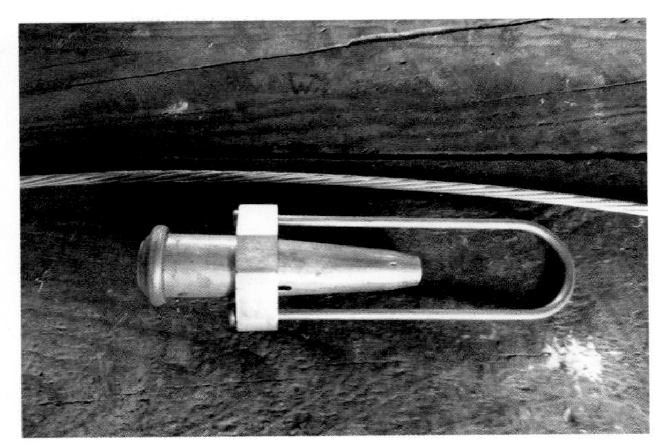

The guy strand tensioner connects the guy wire to the ground anchor. Photograph by Laura Ten Eyck

The guy hook connects the guy wire running from the ground anchor to the top of the end pole. Photograph by Laura Ten Eyck

cylindrical device that the cable runs through that works something like a Chinese finger trap. Once the cable is pulled through the cylinder and drawn taut the cable cannot slip back through.

Guy Hooks. While the ground anchor connects the guy wire to the ground, the guy hook is the piece of hardware that connects the guy wire to the top of the end pole. It is a T-shaped hook with prongs on the back that dig into the pole. It is mounted on the side of the pole facing the ground anchor.

Bolts and Nuts. A long, thick bolt runs through the top of each end pole. The hardware supporting the cable is fastened to either end of the bolt with nuts.

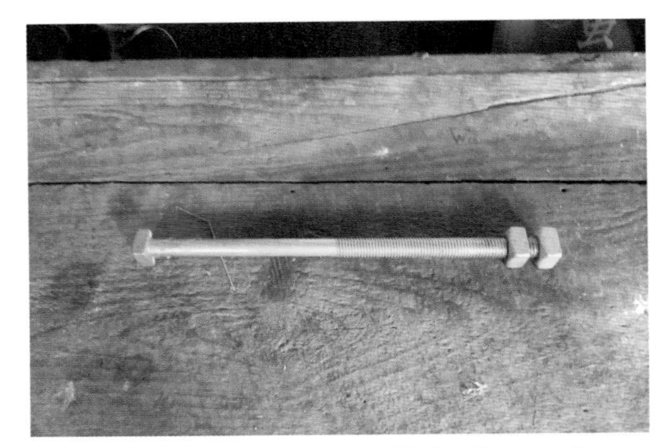

Bolts secured with nuts connect cable hardware to the tops of the poles. Photograph by Laura Ten Eyck

The dead end secures the cable and connects to the hardware on the pole. Photograph by Laura Ten Eyck

The thimble eye nut and plate hardware attaches the cable to the side of the end pole facing into the hop yard. Photograph by Laura Ten Eyck

The two-bolt guy clamps and staples secure the cable to the top of the poles in the hop yard. Photograph by Laura Ten Eyck

Dead ends. Dead ends are long, twisted, U-shaped pieces of metal. Dead ends are used to hook the cable onto the hardware at the top of the pole. The rounded end of the dead end hooks to the hardware on the pole. The cable is then secured by twining the two twisted prongs of the dead end around the end of the cable.

Thimble Eye Nuts and Plates. The thimble eye nut is fastened to the bolt at the top of the end pole on the opposite side of the pole from the guy hook. A plate sits between the thimble eye nut and the pole itself. The cable that runs the length of the hop yard is attached to the thimble eye nut using a dead end, as with the guy wire and the guy hook.

Two-Bolt Guy Clamps and Cable Staples. The cable running the length of the hop yard is attached to the top of each pole with a cable staple. On every fifth pole the cable is secured with a two-bolt guy clamp, which is essentially a rectangular piece of metal divided in half lengthwise. The cable is inserted between the two halves; these are then secured to the top of the pole with nuts and bolts.

Equipment Needs

Whether it is digging deep holes, moving heavy hop poles, or stringing cable at high elevation, erecting a hop yard involves many types of activities that require different types of equipment. Since it is unlikely that you will have all this equipment on your farm, you will have to buy, rent, or borrow certain things. Before you start constructing your hop yard, it is necessary to think the whole process through and make sure you have access to what you need when you need it.

The cost of equipment is one of the most prohibitive factors when starting a commercial hop yard, so renting or borrowing as much as you can is best. But for a commercial hop yard there is one thing you are really going to need throughout the year, and that is a tractor. We have a New Holland TC45 tractor, which we bought used for $16,000. We later came to feel it was too lightweight and therefore unstable when moving heavy loads. To correct this problem we had calcium put in the tires.

Here are some activities you'll have to factor into your equipment planning.

DIGGING HOLES

Although a shovel or manual posthole digger will work for a small number of holes, it is impractical to dig holes for poles in a commercial hop yard by hand. In most cases, a tractor with an auger or posthole digger that can go as deep as 4 feet (1.2 meters) will suffice. However, if you have particularly rocky soil, as we do, you may need to take things a step further. To dig the holes for our hop poles we borrowed an auger, which we used with our tractor, but we also rented a skid steer with a pneumatic auger at a cost of two hundred dollars a day, including delivery and pickup of the machine. We started out using

The auger that we borrowed and used on our tractor was able to do the job of digging 4-foot-deep holes, but it took longer.

The skid steer with a pneumatic auger was a powerful machine that dug deep holes quickly but cost two hundred dollars a day to rent.

The poles we installed in the hop yard were heavy. We experimented with different techniques for moving them with the tractor. Photograph by Laura Ten Eyck

the borrowed auger, but because of a lot of big rocks it was slow going. We decided to try renting the skid steer to see if we could dig more holes deeper and faster. The skid steer was faster, and we felt it was worth the cost to rent it.

MOVING POLES

When your poles are delivered they will be unloaded into a pile. It will be necessary to transport each pole to the location of the hole into which it is to be inserted. These poles are very heavy, and although they can in fact be carried by a team of people, when you are putting up an acre of them, that quickly becomes impractical. A tractor will be necessary. The poles can be chained to a tractor's front-end loader or lifted with the tractor's forks for transport.

GETTING HIGH

When you are erecting your hop yard, sooner or later you are going to have to get high. Elevation is necessary to install hardware and run cables when you are first putting up the hop yard and will continue to be necessary to make repairs. Running string from the wires of the hop yard is an annual activity that involves getting up in the air to attach the strings to the trellis. In days gone by, hop yard workers were known to do this work on stilts.

If your hop yard is small in size, the high elevation work can be accomplished on a tall ladder. When we first got started with our pilot hop yard we purchased a 15-foot (4.6-meter) tripod aluminum orchard ladder. When our linemen neighbors came to help out with stringing the hop yard they nimbly climbed our

Stilts were commonly used in hop yards in days gone by and continue to be used in some European hop yards today. Photograph Courtesy of Rural Life Centre

A 15-foot orchard ladder can be used to access the upper regions of the hop yard.

anchor poles with the use of small blades called gaffs strapped to their boots and hung from the poles by body belts. Such a technique may work great for linemen but is not feasible for laymen.

When it came time to actually run the cable from pole to pole we rented a man lift. There are many different types of man lifts. When considering renting or buying one for high elevation work in a hop yard you need to make sure the platform raises to at least 16 feet (4.9 meters) in height, allowing access to the top of the poles. Stability is another crucial factor. Hop yards are rarely perfectly flat. To work safely at a high elevation you need to be on a stable platform. We rented a Genie man lift that had four-wheel drive at a cost of two hundred dollars a day, including delivery and pickup of the equipment.

Farmers, being farmers, have rigged up homemade versions of the man lift, achieving varying degrees of height and safety. Some have mounted stairs on wagon beds. An apple bin lifted on tractor forks also works. Though we rented the Genie in the spring, by harvest time we had our own homemade man lift made from a metal cage that used to contain a plastic water tank mounted on a pallet. To get high enough to cut the hop bines, we strapped the cage to the forks on the tractor and lifted it up so that our friend in the cage could cut the bines off the wire. Out west when it comes time to string the hop yard,

Workers who climb utility poles for a living donned their gaffs to install hardware on the top of the poles. Photograph by Laura Ten Eyck

a tractor tows a specially built platform that can hold half a dozen men.

RUNNING CABLE

Depending on how much cable you've got, a spool of cable can be pretty big and heavy. When it comes time to run the cable equipment it will be necessary to transport the heavy spool up and down the rows. A tractor with forks worked well for us. We ran a metal pole through the center of the spool and laid it horizontally along the tractor's forks. As the tractor moved forward, guided by people, the cable easily unspooled.

TOOLS

In addition to heavy equipment there are a number of tools involved, and it is good to plan ahead and make sure you have everything on hand. When installing the hop poles you need to have:

a *plumb* on a length of string to make sure the poles are in the ground straight;
a *sledgehammer*, used vertically, to tamp down the soil around the poles once they are erected;
a *power drill* to install the hardware onto the poles, with a *bit* long enough to go through the entire thickness of an end pole and make a hole wide enough in circumference for the bolts you need to install;
a *wrench* to tighten nuts;
a *14" chain saw* to trim a pole that is slightly too tall or too thick;
cable cutters of sufficient strength to cut the thickness of cable you are working with; and
a *come-along cable puller* to tighten cable.

Most of these tools are going to be of ongoing usefulness. Although they can certainly be borrowed it makes sense to purchase them and keep them on hand for the maintenance of your hop yard. However, cable cutters and the come-along cable puller are expensive, and it is possible to rent them.

Human Power

Once you have laid out your hop yard, assembled your materials, lined up your equipment, and gotten your tools together, it is probably more than apparent to you that this is a monumental project and you are going to need some help. You are going to need a lot of friends, perhaps a few professionals, and (in an ideal world) some who are both. One of the things that amazes me the most about growing hops is how many people want to help. Whether it's weeding, picking hops, erecting hop poles, or running cable, since Helderberg Hop Farm's inception we have been assisted by an army of friends, acquaintances, and neighbors—and have even made new friends with complete strangers who have come along and

The Genie man lift we rented provided elevation necessary for reaching the tops of the hop poles to attach hardware and run cable.

To run the cable we put a metal pole through the wooden cable spool and rested it horizontally on the forks of the tractor. Driving the tractor forward allowed the cable to unspool. Photograph by Laura Ten Eyck

A rotating crew of friends helped us install all eighty poles in our 1-acre hop yard. Photograph by Laura Ten Eyck

volunteered to help. Feed these people and give them beer and make sure to let them know how much you appreciate them.

Installing the Poles and Cables

Before any actual work begins it is best to mark the territory. Remember you are going to be assisted in a wide variety of tasks by various people who may have little understanding of the big picture.

Poles form the infrastructure of the hop yard, and identifying the future location of the poles will help give your virtual hop yard shape. Outdoor-grade marking flags are very handy for this purpose. Such a flag is basically made of a metal rod topped with a brightly colored flag that, stuck in the ground, is easily visible. Use one color flag to mark the locations of your ground anchors, another color for your end and side poles, and a third color to mark the location of your inside poles.

In order to determine the exact location of each feature you will have to measure. When we marked out the locations for our inside poles, which are 35 feet (10.7 meters) apart, we used a piece of rope cut to that length, with a metal stake tied to either end. Stick one stake in the ground where your initial pole will be, walk down the row until you run out of rope, and stick the other stake in the ground. Mark the location of each stake with a flag and repeat. One word of caution—make sure not to use a rope with any stretch. One person may pull the rope taut when marking locations (me) and the other may let it lie loose (Dieter), resulting in differing measurements. Once the locations are marked it is time to dig.

We hired a contractor with a boom truck with a pneumatic power auger to dig the holes for our end poles.

The end pole was installed with the use of the boom truck's mounted crane.

To bury our ground anchors we used our tractor with a borrowed auger to drill holes 12 inches (30.5 centimeters) in diameter and 4 feet (1.2 meters) deep. We then widened the hole and laid the ground anchor in at an angle, backfilling by hand. When installing the recycled utility poles that serve as our end poles we had to have some help. We hired an independent contractor who had a boom truck with a pneumatic power auger. He drilled the holes for the end poles, which were 6 feet (1.8 meters) deep and 14 inches (35.6 centimeters) in diameter. He then, one at a time, attached each pole to the boom with a rope and used the boom to lower the pole into the ground.

As mentioned earlier, we have rocky soil, so when it came time to install the row poles we rented a skid steer outfitted with a pneumatic auger to dig the holes, all of which were 4 feet deep. The skid steer is a small, boxy piece of equipment on top of four big tires, within which the driver sits. The front of the machine is outfitted with a pair of arms that raise and lower, to which many different types of attachments can be added. Inside the box the operator controls the machine with a set of joysticks. We had a crew of volunteers lined up for the weekend to install the hop poles. The skid steer was delivered on a Wednesday and we worked feverishly to dig all eighty holes before the weekend. By the end of the day Friday the holes were dug. During the night nearly 2 inches (5.1 centimeters) of rain fell.

When we reported to the hop yard for duty the next morning we found many of our eighty holes filled halfway or more with mud and rocks. Our volunteer pole installation crew included friends with diverse professional qualifications, none of which necessarily prepared them for the job at hand: Kevin, director of a health-care nonprofit; his wife, Camille, then assistant director for the state Committee on Open Government; Mark, director of our regional land trust; Jeff, an IT guy for our county government;

The skid steer with the pneumatic auger proved to be a powerful tool for digging holes in rocky soil. Photograph by Laura Ten Eyck

After many of the holes we dug with the skid steer and pneumatic auger filled in from a rainstorm, we had to borrow the tractor-drawn auger again to dig the holes back out. Photograph by Laura Ten Eyck

Transporting a hop pole chained to the elevated tractor forks proved to be dangerous because the pole swung from side to side. Photograph by Laura Ten Eyck

Chris, a local banker; Stuart, salesman of restaurantware; and Matt and John, both white-collar professionals for the State of New York. But they were strong and enthusiastic, and Camille, thanks to a college job on a tree planting crew, knew how to drive a tractor and use an auger—which, as it turned out, proved essential. She drove the tractor down each row, using the auger to displace the dirt and rocks that had washed back into the hole in the rain.

As Camille excavated the holes, the rest of the crew strategized about the best way to contend with the poles. To move them, we had planned on using a technique we had seen demonstrated in a video from UVM that involved a chain and a tractor with a front-end loader. The basic concept was to attach the looped chain to the tractor's front-end loader and use it like a sling to lift and transport the pole. The trick was to find the pole's middle point so that it could balance in the sling. It is important to emphasize that these 22-foot (6.7-meter) poles are really quite heavy, especially if they are still green and contain a lot of moisture, as our hop poles did. We estimate that each pole weighed approximately 250 pounds (113.4 kilograms).

Our New Holland TC45 is, as tractors go, a fairly lightweight machine—much lighter than the one used in the UVM video. We quickly found out it was not heavy enough to counterbalance the weight of a hop pole when it was lifted with the front forks. That lack of balance coupled with the trouble we were having finding the sweet spot where the pole achieved a steady balance in its chain sling left us in trouble. Imagine an enormous pole dangling several feet above the ground from a chain hanging off the front of an unstable tractor with your friends alternately trying to grab hold of the heavy pole and dodge out of the way as it swings wildly to and fro. It seemed like a good time to break for lunch. We ate pizza and drank a moderate amount of beer while Camille serenaded us on her ukulele.

Plan B, improvised after lunch, was to maneuver the pole into a horizontal position lying across the tractor's forks. In this manner the pole was then transported to an area near its hole. A team of men then lifted the pole horizontally and carried it to the hole, tilting it so that the lower end dipped into the hole—then pushing and shoving with all their might until the pole stood relatively straight in the hole. This method, while good for the building of both team spirit and machismo, soon resulted in strained backs and sheer exhaustion.

We then hit upon Plan C, which was to maneuver the pole as close to the hole as possible with the tractor and forks. Then, with the pole still lying across the raised forks, the guys tipped the end of the pole into the hole and held it steady while the forks raised the upper end of the pole higher into the air, sliding the pole into the ground via the force of gravity.

Even with five men carrying a pole they still suffered back strain. Photograph by Laura Ten Eyck

Once the pole was in the ground we used a plumb line to make sure it was standing straight. A plumb line, also called a plumb bob, is simply a weight on a string. When held up in the air with the weight completely still the string forms a straight vertical line. The person holding the plumb line and the pole in their line of sight directed others to maneuver the pole until it lined up with the string. Once the pole was straight we backfilled the hole with the surrounding dirt and packed it using the top of the head of a sledgehammer as a tamper. It is very important to make sure that the hole is well filled with dirt packed as firmly as possible; it is the compacted earth around the base of the pole that is going to hold it in place. By the end of the day half of the poles were in place.

The next morning, while pole installation continued, our friends Steve and John showed up to

pole at an angle using the forks on the tractor while people guide the end of the pole into the 4-foot-deep
by Laura Ten Eyck

Trish demonstrates how to hold the plumb line to make sure an installed pole is straight.

volunteer their high-wire skills. The two kept up a continuous line of banter and bickering reminiscent of brothers, or perhaps an old married couple.

With their arrival we knew right away we were on a different footing. Anticipating the need for elevation we had rented a man lift. John quickly powered it up, climbed aboard the platform, and raised himself up. Needing additional height, he climbed to the top of the safety cage. Balancing there to investigate some vestigial hardware at the top of the pole left over from its previous life, he realized he had forgotten his hammer and called down to Steve for it. Steve handily tossed an enormous hammer skyward and John, balancing on top of the safety cage's uppermost bar at a height of about 18 feet (5.5 meters), plucked the spinning hammer from the air. Steve, disdaining the man lift, donned his gaffs and speedily climbed the adjacent pole. Leaning back into his body belt, he

Steve attaches the guy wire to the ground anchor with a guy wire tensioner. Photograph by Laura Ten Eyck

The guy wire/ground anchor setup is now complete. Photograph by Laura Ten Eyck

Attaching cable to the tops of the end poles. Photograph by Laura Ten Eyck

Finishing wrapping the cable in the dead end, which is run through the thimble eye nut. Photograph by Laura Ten Eyck

hung in apparent comfort while he installed the bolt that goes through the top of the pole, followed by the guy hook. Steve, too, had left his tools on the ground, but his chain saw and power drill were attached to his waist by long pieces of rope. When it came time to use them he simply pulled them up by the rope hand over hand. When he was done with them, he lowered them back down.

Steve and John worked together to connect guy wires to the tops of the guy hook and the top of the poles. A device called a dead end—which looks like a long, skinny letter *U* made of twisted metal—was hooked to the T-shaped portion of the guy hook. The prongs of the *U* were then wrapped around the end of the cable, securing it. The guy wire was then run down to the ground anchors, pulled tight with the come-along, and fastened to the top of the ground anchors with a guy wire tensioner. The guy wire tensioner's U-shaped rod was threaded through the hole at the top of the ground anchor. The cable was then fed through the cylindrical part of the tensioner, which functions something like a Chinese finger

A metal pole laid across the tractor forks runs through the spool of cable and allows the cable to unspool as the tractor moves down the row. Photograph by Laura Ten Eyck

trap. Once the cable was pulled tight it could not slide back out. With this accomplished it was time to run the cable for the trellis.

On his gaffs Steve attached a thimble eye nut to the bolt on the opposite side of the end pole from the guy hook. He employed another dead end to secure the cable end, this time threading through the eyelet. The initial running of the wire down the first row was done at ground level. With the cable running down from the top of the end pole, we ran a metal rod through the center of the spool of cable and laid it horizontally across the forks of the tractor. The tractor then slowly drove down the row, allowing the cable to feed off the spool onto the ground.

Once the cable was laid on the ground for the full length of the row, we brought the man lift back to the beginning of the row. There we raised the cable, attaching it to the top of the first row pole with a cable staple. As we proceeded down the row with the elevated cable it became quickly apparent that the poles were not all the same height. This was no doubt because the poles were not all buried at the same depth. Apparently we had not been able to fully excavate the holes that had filled in during the rainstorm. This necessitated the painstaking process of measuring each pole with a tape measure, determining the height of the lowest pole, then trimming the tops off all the other poles with a chain saw to that height.

With this accomplished, the hoisting of the cable began anew. We attached the cable to the top of each pole with a cable staple. On every fifth pole the cable was attached using a two-bolt guy clamp for extra strength. Once the end of the row was reached the cable was tightened with a come-along, then secured with a dead end threaded through the eyelet.

We learned to use the man lift and a chain to move and lift the hop poles.

The process was repeated for each row and, thanks to our neighbors, was completed in a day. Fresh crew members kept showing up, too: Trish, a mom and avid organic gardener; Alex, who works for the electric company; Woody, an actual electrician; Andy, an attorney; Kathy, a retired school superintendent; and Joe, an actual farmer, moonlighting after his shift at the orchard. An added benefit to having the linemen there was that they showed us how to raise the poles using the man lift. Because the man lift was able to reach a higher elevation than the tractor with the forks, it was easier for the guys to get the poles installed.

Stringing the Trellis

Once the cable was up, there was one more high-level task to be accomplished—the stringing of the trellis. With all the concern regarding cable and trellising it is sometimes overlooked that the trellis is not what the hop grows on but instead is what supports the strings that the hops climb. Stringing the trellis is an annual task that involves tying strings to the cables and running them down to each hop crown on the ground. We hung on to our rented man lift for one more day to accomplish this task. Because the bines are cut at harvest time, the stringing of the trellis is an annual task we wi[illegible]e to get to a high elevation to accomplish.

The string traditionally used in hop yards is called coir. Coir is a strong, bristly twine from Sri Lanka made of coconut husk. It is preferred by hop growers because its rough surface is easy for the hop bine to climb, it is strong, and it is biodegradable. Coir can be purchased from hop yard suppliers (see Resources). It is usually precut and comes by the bale. A bale of 3,200 strings costs about $400 and a partial bale of 1,000 strings costs about $150.

The coir is tied to the trellis cable with a clove hitch knot. Before stringing the trellis you need to soak the coir in water for twenty-four hours. This makes it more pliable, and as it dries it will shrink,

Coir is the standard material used when stringing hop yards.

Coir, twine made from coconut fiber which the hop bine will climb, is hung from the trellis above each hop plant and anchored at ground level adjacent to the crown. Photograph by Laura Ten Eyck

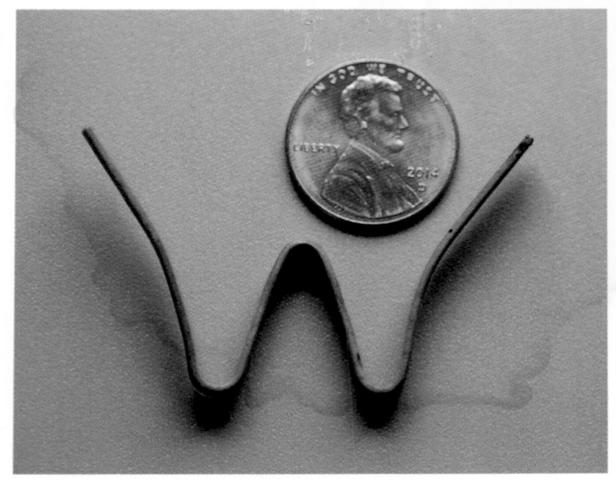

The W clip is a device that anchors the end of the coir below ground.

A finished clove hitch knot (*in the photograph*) and how to tie one (*in the illustration*). Illustration by Dahl Taylor

The W clip holding the end of the coir is inserted into the ground using the W clip applicator.

tightening the knot holding it to the trellis. Once you are ready, secure a length of coir to the trellis above the hop crown.

Once the knot is tied, drop the length of coir to the ground. Then embed it in the ground adjacent to the hop crown by anchoring it with a little metal contraption called a W clip. The W clip is inserted into the ground using a tool called a W clip applicator that has a forked end. You attach the clip to the end of the applicator and place the length of coir that touches the ground into the interior upside down V shape in the center of the W. Place the clip into the prongs of the applicator and insert the tip of the applicator holding the W clip into the ground. Push the W clip into the ground to a depth of about 6 inches (15.2 centimeters) by applying force with your foot, using the horizontal bar above the end of the applicator. The buried clip and end of the coir will anchor the string to the ground, creating a stable vertical structure for the hop to climb. W clips are inexpensive, costing just pennies apiece and coming in quantities of one thousand or five thousand. W clip applicators can be purchased for about sixty dollars. Both can be acquired from hop yard suppliers (see Resources). Once the coir has been secured below ground with a W clip there will still be a length of coir left above ground. This should be cut down to a manageable length so that it does not interfere with weeding and cultivation activities around the crown.

Joe imbeds the W clip holding the coir into the ground adjacent to the hop plant.

Installing Irrigation

The final step in preparing your hop yard is the installation of the irrigation. Since hops do not like their leaves to get wet, sprinklers are not a good option. Drip irrigation, which delivers the water directly to the ground, is the best choice. However, you'll need to take some special precautions with it. Normally drip irrigation is delivered by a hose that lies directly on the ground, emitting water slowly into the soil. But hoses on the ground interfere with hop cultivation, particularly a practice called crowning—the mechanized removal of the top section of the hop crown. Crowning is done in early spring to prune back old growth, remove any early emerging shoots, and control diseases such as downy mildew that are spread through shoots (referred to as basal spikes) that emerge directly from the crown. So drip irrigation that lays directly on the ground must be rolled up at the end of the season to prevent it from getting wound up in the mechanized equipment used for cultivation and crowning.

In very large hop yards this may be impractical. For this reason, in some hop yards, drip irrigation is elevated about 10 inches (25.4 centimeters) off the ground to allow mechanized blades access to the crown. This saves time but also has some drawbacks. The vegetation's exterior at the base of the plant, where disease often begins, will become wet during irrigation, which could lead to an environment favorable to the reproduction of mold spores. Also, because the water will be traveling through the air to a certain extent, during hot weather some evaporation will occur, resulting in water loss that could be a problem in the event of a shortage.

Most commercial hop farms use drip irrigation that lies directly on the ground. The water flowing through this irrigation can come from surface water such as a pond or from a well below ground. The

An example of drip irrigation lines in a hop yard elevated to allow for crowning.

Drip irrigation line on the ground next to a young hop.

system is composed of the water line that brings the water from its source, a filter, and the hose that carries the water to the field. Keep in mind that if you are using surface water as a source for drip irrigation, you will need a good filtration system, because surface water generally has more particles in it and can clog the irrigation system. A section of pipe called a manifold runs horizontally along the end of the field. Lines carrying water run from the manifold down the rows of plants delivering the water.

There are two options for those lines. You can use hoses with external emitters that you can install yourself to release water only at specific points of your choosing, such as at each hop crown. Or you can use drip line with preinstalled emitters that release water at evenly spaced points along the row. Many hop farms also use their irrigation system to fertilize their hop yard, pumping fertilizer into the irrigation water through a fertigator—a system installed between the water line and the manifold that feeds a measured amount of fertilizer into the water.

Installing irrigation costs approximately $1,300 an acre. You can purchase parts and build your own irrigation and fertigation system, which may be cheaper but more time-consuming, or you can hire a company to do it for you. Factors to consider when figuring out what you need include water pressure, control valve placement, tubing size, and methods such as air vents to prevent the system from sucking in soil when the water is turned off. Terrain is also a

The fertigator in our 1-acre hop yard regulates the emission of liquid fertilizers such as manure tea and fish emulsion into the irrigation water. Photograph by Laura Ten Eyck

factor. We could not at first figure out why one end of our hop yard was growing so much slower than the rest of the yard. We finally realized that the ground began to slope up ever so slightly at that point and that the water pressure was not sufficient enough to push the water up the gradation. Another thing to consider in the design of the system is the need to be able to take sections apart to make repairs and unplug clogs, which occurred more frequently than we anticipated.

With the poles, trellis, and irrigation in place you may be ready for a vacation. Of course you won't be able to go anywhere because you will be flat broke. So you might as well turn your attention to planting rhizomes, since everything you need to grow hops is now in place. Installing a hop yard is a lot of work, so much so that many new hop growers can think of nothing else. As overwhelming as the task of installing the infrastructure might seem, it is important to remember that preparing the soil and controlling weeds remains the most important first step in establishing a hop yard. If you make sure to do this before you put those rhizomes in the ground, you will have made life oh so much easier for yourself. Trust us, we know—because, as with everything, we learned the hard way.

Constructing a hop yard is expensive and involves a lot of hard work. Preparing the soil before planting and setting aside plenty of time for advance planning will help control costs and save on labor in the long run.

CHAPTER 6

From the Ground Up, Up, and Up

Mid-July is one of my favorite times to walk in the hop yard. The vigorous green bines climbing up long lengths of twine have easily reached the top of the trellis system. They've had plenty of time to develop the side shoots on which the precious hop flowers form, and those shoots hang in fluid columns, undulating lazily in the breeze like rows of tall, green hula dancers. In what's called the "mid-burr" stage, the developing flowers dangle from the horizontal tendrils in berrylike clusters. Up close they look like little, luminescent, green pinecones. When you pluck one, crush it, and take a whiff, the delicate beginning of that distinct, bitter, hoppy aroma rises from the dry leaves.

You must plant once and tend often to get to this point, and harvest of course is still several weeks away, but it is a nice midpoint to a process that begins, if you are in your first year of growing, very early in the springtime. Here, we'll review what's needed to select and plant your rhizomes, train the bines, and cultivate your hop plants.

Getting Started: Rhizomes

Before you can have hop plants you must have hop rhizomes. A hop rhizome is part of the hop plant's root system. Like a root, a rhizome grows below ground but is actually a sort of subterranean stem sporting buds that will sprout the shoots that work their way to the soil's surface and become the hop bine—as we learned in Chapter 2.

There are two ways to obtain hop rhizomes. One is to buy all the rhizomes you need to start your hop yard and have them shipped to you—the fastest and simplest way but, as with many of life's conveniences, also the most expensive. Hop rhizomes usually sell for between four and five dollars each. Prices vary depending on the seller, the variety, and the size of the rhizome. Many vendors offer volume-based discounts and wholesale pricing, and most start taking orders in January, with discounts sometimes available for ordering early. To fill orders, suppliers dig and ship rhizomes in March. They should be planted

The hop rhizome is an underground stem that grows at the surface of the root system. Hop bine shoots sprout from buds on the rhizome and make their way above the surface of the soil in search of sunlight.

as soon after receipt as possible, but if they can't be planted right away they will stay in relatively good condition if you keep them moist, wrapped in plastic, and refrigerated.

The alternative to purchasing all your rhizomes is to buy only a few rhizomes, get some plants growing, then cut your own rhizomes from them once they are established. You could also get some neighbors growing hops to let you cut rhizomes from their hop crowns, or take cuttings from their plants' shoots, and use these to propagate your own hops. Propagating your own rhizomes and starting plants from cuttings is not complicated but takes time and is a great option if you are not in a big hurry. (It is worth noting here that if you are a person in a big hurry hop farming may not be for you.) Propagating your own hops is also the way to go if rare, heirloom varieties are your focus or you are experimenting with hops you have collected from the wild.

If you buy hop rhizomes it is important to buy them from a good source that is only selling disease-free rhizomes. Fungal diseases such as downy mildew are among the biggest threats to hop yards, and these diseases can be carried in the rhizome. Farmers in the East, establishing hop yards where hops have not been cultivated for a long time, are being encouraged by land grant colleges such as UVM and Cornell to procure hops from sources participating in the United States Department of Agriculture's (USDA) National Clean Plant Network. This publicly funded network of regional facilities maintains disease-free working collections of what the USDA calls "specialty crops." By definition, specialty crops are essentially any food crop that is not considered a commodity, as crops like corn, wheat, or soybeans are. The specialty crops cultivated by the National Clean Plant Network are berries, citrus, tree fruits, grapes, and hops, which were added to the network in 2010. Currently there is only one source of certified disease-free hops, and that is the Clean Plant Center Northwest, located at Washington State University's research and extension center.

A bucket of hop rhizomes heading out to the hop yard to be planted.

These hop cuttings, dipped in rooting hormone and placed in jars of water, have begun to develop roots.

Once hop cuttings develop their roots it is time to plant them into soil so they can continue to grow.

Propagating Hops

You can propagate hops by taking a rhizome cutting from another plant or by taking cuttings from young shoots or sturdy bines.

CUTTING FROM RHIZOMES

Although it is convenient to purchase rhizomes to get started, once your plants are established and you want to expand, you can cut rhizomes from your hop yard to propagate new plants. If you want to save money you can also cut rhizomes from wild plants or another person's planting, but remember that you will have no assurance that these rhizomes will be disease-free. If you are foraging, make sure that the plant you are cutting rhizomes from is female, because the plant that grows from a rhizome is going to be a clone of the original plant.

You must also make sure that the plant you choose to cut rhizomes from is fully mature, as cutting rhizomes from a hop that is not fully established will be too stressful for the plant and stunt its growth. Rhizomes should be harvested in the very early spring as soon as you can dig in the ground—usually March. Cutting the rhizomes is a pretty simple process. Just take a shovel or hand trowel and dig down into the top of the hop crown, which lies just below the surface. The bigger the crown, the more rhizomes you can cut. For example, from a healthy mature crown that is about five years old you can cut as many as fifteen rhizomes of 3 to 4 inches (7.6 to 10.2 centimeters) in length without harming the plant. We have one extremely large hop that is about twenty-five years old, and we have cut as many as ninety rhizomes from it with no ill effect. The hop roots are brown and woody and look kind of like underground dried grapevines. The rhizomes are shoots emerging from the roots. They are pale, almost looking like a thin branch that has been stripped of bark. The buds appearing on the rhizomes are greenish-white or sometimes purple polyplike bumps. Select rhizomes that are not growing on the periphery of the central part of the crown. Using a knife, simply trim the selected rhizomes away from the roots. Cut the rhizome itself, leaving a length of at least 3 inches of rhizome still attached to the root. Then cut the rhizomes into lengths 3 to 4 inches long. Each length should have as many buds on it as possible. These rhizome cuttings, planted directly into the ground in early spring, will send up shoots and produce a new hop plant.

This hop cutting has begun to sprout its own roots and is ready to be planted in soil.

TRANSFORMING SHOOTS INTO RHIZOMES FOR PROPAGATION

Some growers have experimented with propagating new hop plants from sturdy bines growing from the base of the hop plant. In late summer or fall, identify a few shoots growing close to the ground. Dig a narrow trench a few inches deep in the ground, extending out from the crown. Leaving the shoot attached to the crown, bury it in the trench and mark the location. Over the winter it should transform from an aboveground shoot into a rhizome with buds. Early the following spring, dig up the shoot, sever it from the crown, cut it into sections 3 to 4 inches long, and plant the sections as you would a rhizome.

Some growers have followed the same general procedure, but instead of leaving the shoot connected to the crown, they prune shoots in the fall and bury them over the winter. When they dig the shoots up in the spring, they cut them into sections and plant them as they would rhizomes, producing new hop plants.

PROPAGATING HOPS FROM VEGETATIVE CUTTINGS

Vegetative cuttings are different from rhizomes in that the cutting has been taken from the plant vegetation growing aboveground rather than the underground system of rhizomes.

If you order plant cuttings, when they arrive dip the bottom 1 or 2 inches (2.5 or 5.1 centimeters) of each cutting into a rooting hormone solution, then insert the cutting vertically, stem first, into a small pot or flat filled with a moist sterile growing medium—making sure that the bottom node on each cutting is buried. We use a commercial brand of rooting hormone (Rootone) available at most garden centers. Keep the cuttings warm, and do not allow the soil to dry out. If the cutting successfully takes root it will begin to grow. Give it some time to get established, then move the cutting to a larger pot. The potted cutting should get full sunlight but be protected from wind and cold. The soil needs to be kept moist. When the potted cutting begins to grow, fertilize it with diluted manure tea or fish emulsion. When the plant becomes larger and stronger, it can be planted directly into the hop yard.

To take cuttings from existing plants use a sterile knife to cut young shoots from the plant that have three or four sets of leaves. Remove the lower leaves from the cuttings, and place the cuttings in water. Watch for roots to emerge from the cuttings. When roots are visible, remove the cuttings from the water and dip the lower portions with the developing roots into rooting hormone solution. Plant the cuttings in small pots or a flat of sterile soil. Once the cuttings begin to grow, transplant them into their own pots and proceed as above.

Rooted cuttings can be potted and kept in a plant nursery where they can be sheltered and receive care until they grow large enough to be planted in the field.

Newly potted hops grown from cuttings are kept in a plant nursery. Potted plants can dry out quickly, and hop plants grown from cuttings are fragile when they are young. It is important not to let them dry out.

A hop shoot emerging through wood chip mulch in our pilot hop yard in early spring.

In 2011 the Northeast Hop Alliance, an organization supporting hop production in New York and New England and advised by Cornell and UVM, began working with greenhouses in New York to root hop cuttings imported from the Clean Plant Center Northwest and grow stock plants. Plants grown from rooted cuttings of these stock plants are being distributed to growers in the Northeast. These plants are currently available only in limited numbers, but the ultimate goal is to produce enough to fully meet the demand of the region's growing industry.

Going through the trouble of obtaining certified disease-free hops when you are starting your hop yard has many long-term advantages. The rhizomes are easier to propagate. The plants will provide a greater yield. And the resulting hop flowers will be of higher quality. In addition, starting with plants obtained from the National Clean Plant Network is better for the environment because it reduces growers' risk of needing to resort to toxic chemicals to treat outbreaks of disease.

Unfortunately rhizomes are not available through the National Clean Plant Network, only vegetative cuttings. Because of this some growers will only plant rooted vegetative cuttings in their hop yard. Procuring rhizomes for our own commercial hop yard, we have tried a little bit of everything. Before we began scaling up, we had grown hops on a hobby scale for decades and had several well-established Cascade hops growing in the garden and around our farm buildings. We took cuttings from their rhizomes, which worked well, as we knew these were healthy plants. When we decided to establish a commercial hop yard, we started off fairly small by ordering one hundred rhizomes. This was a risk we took, and the rhizomes proved healthy. We started with Brewer's

Gold, an older variety known to have done well in upstate New York, and Centennial, which is particularly disease resistant. In addition we had a number of potted heirloom plants—a local version of Cluster—that our neighbor had been safeguarding for thirty years and bequeathed to us.

The following year we propagated soft-tissue cuttings from these heirloom plants, which we named the Helderberg Hop after the mountain on which they were found growing. About half of the one hundred cuttings we rooted and potted survived. We did not consider that too bad considering that we did not have a greenhouse. The survivors were planted in the hop yard. We also dug about one hundred rhizomes from a well-established Cascade hop plant we had growing in the garden. From these rhizomes we grew about one hundred plants in pots, which we also planted in the hop yard.

In our third spring, we dug one hundred rhizomes from the hops now established in our hop yard and ordered three hundred more rhizomes. At that point, we had been working with Brewer's Gold, Centennial, Cascade, and the Helderberg Hop. When we purchased new rhizomes, we added Nugget to our mix; we select varieties for their brewing attributes as well as their disease resistance.

This diversified approach to acquiring hops has real advantages. Not only will experimentation help you learn, but obtaining and propagating hop plants from a variety of sources, especially when first starting a hop yard, minimizes risk. If there is a problem with one source and the plants do not survive, you will have an array of plants procured from other sources to rely on. It is not uncommon to hear of farmers who have started their hop yard by ordering all of their rhizomes from one source, only to have found out, when none of them produced plants, that there had been a problem with the handling of the rhizomes at the source. This is a huge setback for a hop yard just starting up because hop rhizomes can really only be purchased commercially in the spring; if you have a total loss you have to wait until next year to order again. A year is a lot of time to lose, especially since hop plants take three years to reach full production. Diversifying your sourcing is important, but nothing is more critical than being sure to start with cuttings and rhizomes known to be from disease-free plants.

A hop shoot springing up through black plastic mulch in our 1-acre (0.4 hectare) hop yard.

Preparing to Plant

Many hop growers choose to erect the hop yard poles and the trellis system on which the hop bines will grow before planting the rhizomes in the ground. That said, it is possible to let hop bines grow in their first year without support. In fact, there is a school of thought that advises against pruning and training the bine onto a trellis system in the first year and instead advocates letting the bines grow on the ground. The thinking behind this approach is that the main goal in the first year

in the fall and let it break down over the winter, you will be all ready for spring planting.

While we will deal with weeds in Chapter 10, it's worth noting here that once the hops are planted, we keep the cover crop growing in between the rows and rely on mulch to keep down the weeds around the base of the hop plants. Apply the mulch at a depth of 3 inches (7.6 centimeters) as soon as possible after the rhizomes are planted to prevent weeds from getting a head start. The aggressive hop shoots will have no problem making their way through the mulch. Mulch can be composed of wood chips, straw, or compost. Nonorganic mulches such as plastic or landscape cloth can also be used. Of course weeds will still grow up through the mulch and cover crop. The trick is to cut the weeds before they go to seed. The aisles can be mowed. For weeds that inevitably make their way through the mulch around the hop plants, a weed whacker intended for farm use comes in handy—but be careful not to cut the hop bines! For weeding in close vicinity of the bines, the only safe thing to do is pull the weeds by hand.

Planting

Whether you're getting peas or rhizomes into the ground, there is something so pleasant about early spring planting. It is uplifting to finally be working outside comfortably after a long winter. The weather is not yet hot, the ground is free of weeds, and there are no bugs. Life is simple. You are planting something in the soil, and you have every reason to believe it will grow. It seems like nothing stands between you and the eventual harvest.

If you have been able to prepare, when you head out into the hop yard in early spring the ground should be relatively weed-free, for the time being anyway. Assuming you have erected your hop yard trellis system before planting, as we recommend, your target area at this time is the strip of ground running between the hop poles. The hop rhizomes will be planted here in rows. If you have chosen not to erect your trellis system prior to planting you should have it laid out in your mind and preferably on paper, too, so that you will know the location of your rows. Better yet, you will have put in survey flags to mark the eventual location of your poles. But trellised or not, your yard should have its drip irrigation system installed by now, with the hoses running along the surface of the ground in each row where the hops will be planted. It is possible to install the irrigation after planting, but you will have to do it very quickly so as not to risk the ground drying out and your rhizomes getting thirsty. In this time of weather weirding, we can no longer count on April showers.

All these issues aside, at planting time it makes the most sense to focus your attention only on the location in which you intend to plant each rhizome. You can worry about the rest of the hop yard once those rhizomes are in the ground.

Removing any remaining weeds in the area in which each rhizome will be planted can be done without too much trouble by hand or with a hoe. Since you are working relatively small areas it is a good idea to take the time to completely remove the weeds from the dirt, roots and all, rather than just chopping them up and leaving them in the dirt, which gives them a chance to regenerate. Loosen about a square foot of soil at each planting location and work in well-composted manure. Some people recommend planting two rhizomes in each location as a safety measure in case one doesn't grow. In the Northwest they put in as many as four to seven!. We have had good success planting just one rhizome in each hole. Each planted location is referred to as a hill. When you first plant, it won't necessarily look like a hill; but once the hop is fully established, the root base (or crown) will form somewhat of a mound about 1 foot (0.3 meter) in diameter.

Leave 3 feet (0.9 meter) of space between each hill. When you place the rhizome in the ground, make sure as many of its buds as possible face upward. These buds are the points from which the

When planting, lay the hop rhizome horizontally in the ground with as many buds as possible facing upward and cover with a couple of inches of soil.

shoots will emerge, so it is important that the shoots start out heading in the right direction—toward the sun. Cover each rhizome with a couple of inches of dirt. Tamp down the soil gently but firmly. Side-dress the ground above the planted rhizomes with a couple of inches of compost lying on the surface of the soil.

If you are planting more than one variety of hops in your hop yard, it is a good idea to label each planting. When it comes time to harvest it is very important that you keep the different types of hop cones separate, as each variety has its own attributes for brewing, and this is how you will market your product.

Water and Mulch

After planting, water the rhizomes enough to keep the ground moist but not wet. How much watering is necessary is going to depend on the soil's ability to retain moisture, the air temperature, and the humidity level. The only sure thing is that too much water will cause the rhizome to rot. The best approach is to keep checking the soil and to water lightly but frequently. With hops it is better to err on the side of too little water rather than too much.

Mulching the planted rows has a number of benefits. A good layer of mulch on top of the planted rhizomes will keep the ground from drying out, protect the emerging hop shoots from frost, and prevent weed growth. You can use any number of types of mulch. We have used straw or composted bark. Mixing some compost in with the mulch also provides a source of slow-release fertilizer as the nutrients leach through the mulch over time. In our second hop yard we covered the rows with black plastic, with

A baby hop plant pokes up from beneath black plastic mulch.

the drip irrigation hoses running underneath. The black plastic works well for weed suppression for the first year, allowing the young hops to get a good head start on the weeds. It degrades and develops tears through which weeds emerge in the second season and is pretty useless by the third year, when you have to layer mulch over the top. After experimenting with various options in hop yards of ½ acre (0.2 hectare) or larger we recommend using a mulch of composted wood chips or bark in concert with the application of an organic herbicide approved for use on certified organic farms, such as Avenger.

Tending the First Shoots

The shoots sent out by the rhizome will work their way up through the ground, then the mulch, soon making their appearance on the surface. One of the risks during this time is frost. In general established hop plants are quite hardy and the vegetation can take temperatures dipping into the 20s Fahrenheit (–6.7 to –1.7 degrees Celsius). New shoots, however, are more delicate. Mulch offers early protection, but once the hop shoot penetrates the surface of the mulch it is exposed to the elements.

The spring that our friend Erica planted her hop yard, she feared frost one night. She went out and bought one hundred 20-ounce, red Solo drinking cups. The people at the store thought she was throwing a keg party, but she went home and placed one cup over each of her baby hop plants, and they were safe and sound that night. She removed them the next morning, and the baby hop plants were good to go.

If hop shoots do get zapped by frost in spring they are unlikely to continue to grow. Although this will slow down production in the yard, it is not the end of the world as the plant solves the problem itself by simply sending up more shoots. You can clip the frosted tip off a damaged shoot but it probably won't help. Sometimes frost-damaged shoots will send out side shoots to compensate, and this is not good for plant productivity. If you do experience frost damage, your best bet is to wait for new shoots to become established, then prune away the frost-damaged shoots.

Training

If you choose to let your hop bines do their own thing the first year, with the focus being on the development of the root system, you'll have bines growing freely on the ground in year one and will begin training once you have erected your system in year two. While it is possible, it is not practical to harvest the flowers from untrained bines on the ground. In fact it is better not to harvest hop cones from a first-year plant. Leaving the cones reduces stress on the plant, allowing it to put all of its energy into growth.

But most small-scale growers—particularly those who plan to brew their own beer with their hops—don't want to miss out on the excitement of the first year's harvest. For these growers, and those that want to better control weeds and minimize disease risks, it is important to get the hop vines up off the ground and growing vertically. Once the shoots are established, it is time to start training.

Hop bines will climb anything they can get to. Unlike a typical vine that sends out tendrils that it uses to grasp when it climbs, the hop raises itself using its trichomes, the prickly hairs that grow on the hop plant's stem and the undersides of its leaves. Operating sort of like Velcro, the trichomes allow the plant to get a grip and work best on surfaces that are rough in nature. This is important to remember when providing your hop plants something to climb. If you are using a traditional trellis system, described in the previous chapter, this climbing surface will be the coir (our choice) or similar rough, biodegradable rope that you strung from the top of the trellis and anchored in the ground. Each hop plant climbs its string as it grows.

Before beginning to train the bines up the string, allow the shoots to grow to a length of 6 to 8 inches (15.2 to 20.3 centimeters), then select approximately three of the strongest, healthiest shoots and prune the others away. Do not choose the first shoots to emerge; they are generally not the strongest. But don't wait too long to prune and train, as the shoots will inevitably begin to get tangled up; then you risk damaging them while trying to untangle them.

When the hop shoots first emerge from the ground they are a reddish-purple hue. As they grow, they turn green with reddish streaks. Select the bines that are the biggest and appear to be the most robust.

Trish, a friend who helps out in our hop yard, trains a hop bine to climb by carefully twining it around the coir.

A trained hop bine continues to climb the coir on its own.

Give preference to bines that are growing from the central part of the crown, closest to where the string is anchored. Take one shoot at a time and, using your hands, very gently and slowly twine the shoot around the string in a clockwise motion. It is important not to damage the shoots when you are doing this because if you do they will stop growing, and this will slow down the plant's development. The trichomes on the hop bines grasp the rough fiber of the coir readily.

Once you have twined the selected bines, prune the remaining shoots back to the ground. As new shoots emerge, continue to prune them back. The idea here is to channel all the plant's energy into the selected bines growing up the string. Some growers choose to let a few shoots continue to grow along the ground as a sort of insurance policy in case something happens to one of the bines climbing the string. Once, when our goats were running loose, the buck came into the hop yard and bit clear through one of the hop bines, killing it. If you have a ready replacement waiting on the ground that you can start training, such a mishap won't set you back.

The hop rhizome will put out a great number of shoots for a couple of weeks in the spring, and you must continue to aggressively prune them back. If you are looking for more hops to plant, sell, or give to your friends, cuttings from these pruned shoots can be rooted. You can also eat pruned hop shoots. Locavores covet any type of edible green sprouting from the earth in early spring, and hop shoots are no different. In fact, they are sometimes referred to as "poor man's asparagus." If you plan to eat them make sure to select only shoots that are young and tender. When raw, hop shoots are on the bitter side and not great in a salad—but they are delicious sautéed in olive oil or butter.

These first-year hops on Goschie Farms in Oregon are climbing the coir. Note the row cover has not yet been planted nor have the hops been mulched.

Recipe for Hop Asparagus

Whether in ancient times or the modern day, when winter recedes people crave fresh greens. I remember visiting Tatarstan early one spring, on a tour of Russian farms, and we dined at a Tatar home. Although there was no shortage of food, an enormous platter heaped with fresh, uncooked onion tops took center stage. People love to celebrate the first tender shoots of the season, and in many places hops are one of the first plants to poke up through the ground. Edible and delicious, young hop shoots were a common spring dish in England, often referred to as "poor man's asparagus." The great thing about eating hop shoots is that this harvest is a by-product of something a hop grower needs to do anyway—pruning spring shoots.

Hop shoots for the dinner table should be cut in April and May. Later in the spring they are too bitter. If you are a forager, young shoots can be cut from wild hops—but shoots from the hop yard will do just as well. Just make sure they have not been sprayed. Some people add the raw shoots to a fresh salad, but even young shoots may be too bitter for the taste of many. Lightly cooking them is a more popular option. The shoots will be purplish green when you cut them but will turn bright green after cooking.

To prepare the hop shoots, brown a bit of crushed garlic in a pan with butter or olive oil, then simply add the hop shoots. Sauté them for just two or three minutes on medium heat until tender. Add salt and pepper, and serve them as a side dish drizzled with the pan drippings. Another option is to just steam the shoots lightly and dress them in lemon juice, salt, and pepper.

Young hop shoots not selected for training are cut and sautéed as a spring green. Photograph courtesy of Kevin Powers of Powers Farm & Brewery (Fauquier County, Virginia)

Once you have selected the shoots you are going to train to climb the coir, you must continue to cut back all the other shoots that emerge so that the plant puts its energy into growing the shoots you have chosen. Photograph by Kevin Powers of Powers Farm & Brewery (Fauquier County, Virginia)

Cultivation

After the initial spring rush, shoots will continue to come up periodically. You will have to prune them as needed, but this will not take up a lot of your time. The hops' new focus will be on climbing as high as they can. Once they have climbed to a height of 4 to 6 feet (1.2 to 1.8 meters), it is time to begin removing the leaves at the base of the bines. Removing these leaves makes it harder for disease and insects to travel from the surface of the ground to the plant. Start this process slowly, and remove leaves over time from the ground up. You will shock the plant if you remove more than three or four leaves at once; try to strip only three or four leaves from the plant each day. As the plant grows, continue to remove more leaves until the lower portion of the bine is completely leafless to a height of 3 or 4 feet (0.9 to 1.2 meters). This is a good time to side-dress the hops again with compost. You'll also need to water regularly through drip irrigation, as well as scouting for disease and insects.

Once the hop bines reach the top of the trellis and start to put out horizontal shoots, these shoots will form buds that grow into hop flowers. The buds start out as tiny, bright-green beads called

When the hop bine reaches the top of the trellis in late June to mid-July it begins to send out sidearms on which the hop flowers will form. At their earliest stage the flowers are called burrs.

burrs, which grow into papery, pinecone-shaped flowers called cones. Throughout the course of the summer you will check the development of these flowers eagerly. The vision of the fragrant harvest of these sticky, brilliant-green hop cones at summer's end provides inspiration while you are going about chores in the hop yard during the remainder of the growing season.

PART III
TENDING

This Cascade hop growing in our garden is twenty-eight years old. It was originally planted by the barn, then transplanted to the garden next to the house, and ultimately to the vegetable garden where it grows today.

brings will do a lot to keep your hops healthy. But you can't get around the fact that in a hop yard there are a lot of hop plants in one place and this is going to create an ideal environment for the pests that like to eat them.

It is often said that monocultures are always manmade and do not exist in nature, but this is not necessarily the case. Sometimes in nature, under specific conditions, one species of animal or plant will take over a place. Think huge stands of reeds waving in a breezy marshland. Or savannahs where one kind of grass, such as wild sorghum, overpowers the others. When conditions change and the population becomes unsustainable, it crashes—and that's okay because it is just nature taking its course. In the wild, other species arrive over time to fill the niche and heal the ecosystem. But if you're selling what you grow, you need to make sure it can keep growing.

Monocultures attract insects and diseases that thrive on that particular crop because of the plentiful food source. When they arrive, they can reach epic proportions and destroy your entire crop. As a farmer trying to make a living, you really do not want nature to take its course because it means you will lose everything you have worked for. Although sometimes it doesn't seem like it, people are part of nature, too, and it is natural to want to defend your livelihood. But in doing so, humans have made some mistakes—big ones.

Fortunately, though, we are finally coming to realize that simply blasting insects and diseases with highly toxic chemicals does not work in the long run. By applying an abundance of toxic chemicals to our crops we

First-year hops in our pilot hop yard nearing the top of the trellis in mid-summer.

not only pollute our environment, damaging the health of numerous life forms, including our own, but we also improve the resiliency of the very species we are trying to kill. When we plant a crop that a particular disease or insect likes, then spray that crop with a toxic chemical to destroy that insect or pathogen, we create an environment in which that pathogen or insect is highly motivated to develop a resistance to the chemical so that it can keep taking advantage of the food source.

I am familiar with the use of chemicals in agriculture. I mentioned earlier that my father, a commercial apple grower, was educated at Cornell University in the 1950s, during the heyday of post–World War II synthetic pesticide use. Spraying toxic chemicals based on the calendar was the means for controlling pests and disease in orchards. I was born in September of 1962, when biologist and author Rachel Carson published *Silent Spring*. It took some time for her ideas to percolate.

As a child, during the 1960s I was lulled to sleep at night by the dull roar of the sprayer in the distance. I lived through battles of man against fire blight raging in the orchard, during which the entire crop was at stake. When a tremendous outbreak of tent caterpillars was chewing up all the leaves on the trees in the woods by my house, we sprayed the woods. As a child playing in the orchard I have been sprinkled with a fine rain of chemicals from the passing sprayer more than once. And Father would be pretty shocked to learn that there were times I actually played in the powdery interior of the building where the bags of pesticides were stored.

But just as my father was on the forefront of agricultural science in the 1950s, he remained at the forefront as time passed, and our farm was one of the first to adopt Integrated Pest Management (IPM) practices in the orchard. As insects and disease-bearing pathogens began to develop resistance to the barrage of chemicals, farmers and scientists started to realize that continuous spraying did not work. In fact, it made things worse. IPM is a management system that was built to respond to the development of spray-resistant pests and pathogens. Instead of relying on a calendar-based spray program, farmers employing IPM now scout their fields for insects and disease, only spraying when their presence reaches a level that will cause a financial loss in the crop outweighing the cost of labor and chemicals to spray. In addition, the types of chemicals that do get sprayed must be rotated so that the insects and pathogens are less likely to develop resistance to any given chemical. There is also a preference for chemicals that target only the specific species doing damage and do not kill beneficial insects.

Let's be clear: IPM practices are in no way "organic" farming, but they are a big improvement over the previous system when it comes to environmental impact, plus they work better. Since adopting IPM practices, my father has taken it one step further and is a now a certified Eco Apple grower. Eco Apple certification requires growers to implement a rigorous protocol to grow apples sustainably in the Northeast; protect pollinators and beneficial insects; and keep air, water, and soil clean by relying on natural pest control methods.

One of the reasons the Eco Apple program, which has its roots in IPM and family farming, was developed is that apples are one of the most difficult crops to grow on a commercial scale using certified organic methods, particularly in the Northeast. The same may be true for hops. My husband and I came to hop growing as organic gardeners who dabbled in small-scale commercial agriculture. For a time we were selling vegetables to restaurants. At another time we raised sheep for meat and wool. But those ventures were small, and organic principles were relatively easily applied. Located by necessity in the middle of an apple orchard that will spray synthetic chemicals when it has to, our hop yard will never be certified organic, but we are striving to raise hops using sustainable practices that do not harm the environment.

That being said, we are also hyperaware that hop production in our region was completely destroyed a century ago by downy mildew and insects. Another concern is that to produce a commercially viable crop hops need to receive a lot of nitrogen within a fairly narrow window of time. Meeting hops' nitrogen needs purely through the application of compost is our goal, and as I write this we have so far achieved it, but it is challenging. We realize there are some circumstances in which we may have to resort to nonorganic materials to produce our crop.

The Quest for Organic Examples

It is possible to grow hops organically, though, and people like Gayle Goschie of Goschie Farms in Silverton, Oregon, have done it. Goschie's parents bought her family's hop farm in the Willamette Valley in Oregon in the 1940s. Today she operates the 1,000-acre (405-hectare) farm with her two brothers, Glenn and Gordon, where they grow grass seed, corn, wheat, wine grapes, and other crops. They used to grow 500 acres (202.3 hectares) of organic hops, too, until recently, when the borrowed machinery they used to harvest their organic hops was reclaimed by its owner. The organic hops have now been merged into the rest of the hop yard and all the hops are raised using conventional practices.

Researchers such as Heather Darby at UVM are trying to find the right organic management techniques for Northeast growers, but it is still early in the process. So when visiting hop farms in the Yakima Valley to get a better handle on some of our pest and disease challenges, Dieter and I ventured to Goschie's

Gayle Goschie of Goschie Farms monitors the hop drying kiln.

farm to see what we could learn. A tall woman with short gray hair, a set of brightly colored earplugs hanging around her neck, and a baby blue iPhone on her hip, she gave us the facts as we stood around her hop kiln. It is quite possible to grow hops organically, she told us, but because it is not possible to deliver the same amount of nitrogen via organic matter as you can with synthetic fertilizer you simply need to accept the fact that the plants will not be as big. "We used emulsified fish to fertilize the organic hops," Goschie explained. "It is difficult to deliver the same amounts of nitrogen with organic formulations to a very hungry hop plant. Organic certification inspectors are inspecting formulations of nitrogen applied during the growing season. This is to ensure nonregistered organic formulations are not being used."

With less nitrogen and thus smaller plants, you will have a lighter bine. So one of the adjustments Goschie Farms made for its organic hops was in trellising and spacing the plants. They used a trellis system only 12 to 15 feet (3.7 to 4.6 meters) tall with plants spaced 2 feet (0.6 meter) apart, as opposed to 3 feet apart. "With less foliage more sun gets in," said Goschie. "By planting at a higher density you may get the same yield per acre even though you are applying less nitrogen."

It gets harder to apply such tricks of the trade when you are dealing with one of the hop's worst enemies—downy mildew. Growers in Yakima, where it is hot and dry, are not threatened by downy mildew to the same extent as those in moister regions. They told us we would learn more about downy mildew in Oregon's Willamette Valley, and when we drove into that valley we definitely saw more green than in Yakima. We drove through enormous nurseries offering an array of ornamental trees, presumably destined for Portland. We saw Christmas tree farms, vineyards, grain crops, and hazelnut orchards. We also saw huge brown dust clouds on the horizon. The first time we saw such a cloud over a vast field of plowed earth we thought it was caused by an isolated wind dervish, but we soon realized the brown clouds rising into the blue sky were caused by tractors plowing the fields—not a sight you would normally see in the East. The Willamette Valley may get more rain than Yakima, but we had a feeling the growers there may not know the meaning of the word wet. We confirmed then what we suspected: growers involved in the Eastern hops renaissance can't necessarily look to the nation's hop-growing mecca for transferable advice.

Still, Yakima and Willamette Valley growers did have some helpful advice. "Moisture will be a problem for you in New York come harvest time," Goschie warned us. "It is one thing to have downy mildew in the spring. It will stunt the growth of the plant. But at harvest time it is deadly because it gets into the cones." We found that out firsthand when we returned home from our trip. Downy mildew had been slowly creeping up on us all summer. We sprayed the hops with copper, which is approved for organic use, but we didn't start soon enough and didn't spray often

Centennial hop cones in our 1-acre (0.4-hectare) hop yard in July. Brown spots on hop cones and damaged leaves are wear and tear from exposure to insects and disease.

View of some of the Yakima Valley's extensive hop yards.

enough. We did not stop it from going systemic. What started as a few blasted leaves here and there ended up in the cones of a couple of the earlier-maturing varieties. These cones were spotted brown on the outside and desiccated on the inside. We salvaged what we could, but for certain varieties it was a major loss.

Growing hops organically means living with a certain level of disease and insect damage, but even in conventional hop yards control of diseases and pests is an issue of management—not eradication.

The Disease Triangle

"Hops breathe," mused Hopsteiner's Nicholi Pitra in a Yakima Valley hop yard. "This creates a humid microclimate conducive to disease. Disease is hard for any monoculture." When it comes to disease, think in terms of the disease triangle. This is Plant Pathology 101. The three interconnecting points that form the disease triangle are (1) the agent of the disease, (2) a susceptible host, and (3) an environment conducive to the development of the disease. You can eliminate the disease afflicting your crop by removing any one of these three elements. It is more likely, though, that you'll be able to simply control it, and you can do that by altering all of the three points to a certain degree simultaneously.

Take downy mildew, for example. Downy mildew is caused by a spore that reproduces quickly under wet conditions in a particular temperature range. To control downy mildew, address the issue of the susceptible host by planting a hop variety that is resistant to or tolerant of downy mildew. Minimize moisture in the environment by locating your hop yard in a well-drained area where there is good air circulation and sun exposure, and use drip irrigation to keep the plants from getting wet. Then watch the weather forecast. When the weatherman tells you that tomorrow it is going to rain and the temperature is going to be between 64 and 76 degrees Fahrenheit (17.8 and 24.4 degrees Celsius) (the optimum temperature range for downy mildew), go out and spray your hop yard with an organic fungicide such as copper before

that rain begins in order to control the downy mildew spores that are waiting for their favorite weather. To prevent the pathogen from developing resistance it is important to rotate fungicides. If you keep your disease triangle in mind and make an effort on all three fronts at once, you might just stand a chance.

Integrated Pest Management

The challenges of raising hops have led many growers to adopt IPM, versus a strict organic protocol, when it comes to handling insects and disease. The basic premise behind IPM is that instead of spending time and money on maintaining a regular spray program you spend time walking in your hop yard counting bugs that eat hops. You can count bugs on leaves or you can set traps and see how many bugs you catch. You don't spray pesticide to kill them until the number of insects reaches a level that will cause you to lose money from the damage they will cause. This is called the "economic threshold." The premise is that there will always be some insects in your hop yard eating the plants, but you don't do anything about it until you have no choice. When you use the IPM system, you use less pesticide—which is better for the environment, your beneficial insects, and your wallet.

IPM is pretty much standard practice in commercial hop yards. The *Field Guide for Integrated Pest Management in Hops*—produced in 2009 by Oregon State University, the University of Idaho, the USDA Agricultural Research Service, and Washington State University—has long been a tool of the trade. In the summer of 2014 Cornell University hired a professional hop scout named Jason Townsend to visit hop yards looking for trouble. Townsend visited our 1-acre (0.4 hectare) hop yard once a month with his magnification gear and dispatched e-mails giving updates on what he was seeing and making recommendations.

In June he reported, "I found the first two-spotted spider mites of the season on June 20. They were present at low levels and in only one yard so far, but

Lily Calderwood, a graduate student at the University of Vermont, checks a bug trap in the trial hop yard.

A downy mildew spike spotted in our hop yard by Jason Townsend while scouting. Photograph courtesy of Jason Townsend

mites can flare up quickly so growers should keep an eye (and magnifying lens) out for them. I saw potato leaf hoppers in much greater numbers and also continued to see low levels of Lepidoptera larvae and some flea beetles."

In his July hop scout report, he wrote, "Downy mildew continues to be a problem around the state with some yards seeing substantial pressure. Pruning new shoots and lower leaves is important to keeping the disease from spreading upward through the bines. Many growers are now on a regular fungal spray program."

Start Healthy, Stay Healthy

The best way to stay healthy is to start healthy, which is why purchasing plants from the National Clean Plant Network, as described in Chapter 6, is an excellent precaution. In fact, the network's motto is, "Start Clean, Stay Clean." In 2010, frightened by the prospect of hop stunt virus, the hop industry joined the network, which was initially started by the grape and fruit tree industry.

Certified clean hop plants are seasonally available through the Northeast Hop Alliance, in cooperation with Zerrillo Greenhouses (see Resources), located outside Syracuse, New York. These are second-generation hops grown from plant cuttings that have tested negative for viruses in a laboratory. These plants are in 4½-inch pots and are sold in trays containing fifteen plants, starting at $67.50 per tray, or about $4.50 a plant.

Choosing the Right Hop Varieties

Beyond making sure your stock is free of disease, it is also crucial to choose varieties that are less likely to succumb to disease and pests. Some of the older hop varieties come with natural resistance to certain threats. Many new varieties are being bred to be resistant to disease and insects. Unfortunately most of these varieties are resistant to one type of threat but susceptible to another. Another problem is that some of the hop varieties your brewer customers favor may not be resistant to a particular threat present in your hop yard. It is important to consider your customers' preferences, and if they coincide with a variety you think you can grow, so much the better. However, because we are new to hop farming in the Northeast and will face many challenges along the way, we think it is most important to focus on varieties we can grow well first so that we can produce the highest quality crop we can, then market that to our customers.

In choosing what varieties we can grow best, we have to look at what our biggest threats are. This is difficult because it can vary from year to year depending on weather conditions and other factors that impact insect populations. One of the real constants in the East is high humidity, and that poses the biggest threat to hops because it is conducive to downy

Healthy plants propagated from cuttings from the National Clean Plant Network can be rooted and grown into healthy hop plants.

mildew. As mentioned earlier in this book, it is essential for hop growers in the East to plant varieties that are less susceptible to downy mildew. There are several sources with recommendations of the best varieties to plant to withstand downy mildew; however, the recommendations do not always match. And some varieties are more susceptible to downy mildew afflicting certain parts of the plant—such as the crown, shoots, leaves, or cones. (See Chapter 8 for more information on disease-resistant varieties.)

It is worth noting that no variety of hops is completely resistant to downy mildew, and this may actually be a good thing. What we have are varieties that are tolerant of downy mildew. There is a big difference. When a variety is tolerant of a disease it means that the disease exists within the crop, but at a manageable level. Resistance allows no presence of the disease at all. If a hop yard were to be planted with only varieties that are resistant to a certain disease, this would mean that the disease could not exist in that hop yard at all. This puts a kind of pressure on the pathogen. Depending on the biology of the pathogen, this complete resistance can result in the eradication of the pathogen or it can force the pathogen to transform itself so that it can live in its 'esired environment. "If you corner a disease it ight change itself," said Pitra. "If you can get a iety to tolerate a pathogen, the disease will not be armful."

ermining whether ien to Spray

at land grant universities have developed ams and disease prediction models to ers what to spray on their crops and v it, based on levels of insects and dis- her patterns. The spray programs for ped in the Northwest by institutions gton State University. But in the s have not been produced on a commercial scale for so long, no such programs or models exist at the time of this writing. In the meantime, Eastern growers can use the University of Vermont's fact sheets on hops and the *Cornell Integrated Hops Production Guide* (see Resources) and make decisions based on what they learn, combined with their observations in the field and the weather forecast. When insects and disease make their presence known in the hop yard, the level of damage they can inflict will be determined largely by the weather, particularly temperature and moisture levels. For example, downy mildew spreads when conditions are wet and temperatures are in the low 60s Fahrenheit (15.6 to 17.8 degrees Celsius). You need to know with accuracy what the weather is at any given moment, as well as what it is going to be. It is worth investing in a weather station that will accurately measure temperature and wind speed and has a rain gauge. In addition subscriptions to weather forecasts specially for farmers can provide you with specific information that the weatherman on the local news won't.

If environmental conditions become conducive to an outbreak of disease or a pest population explosion in the hop yard, it may be time to think about spraying a pesticide. The question is what to spray. Spraying is really complicated. Different types of pesticides are designed to kill different types of pests. For example, fungicides kill fungal pathogens, insecticides kill bugs, herbicides kill weeds. Within each category there are numerous products. Some of these pesticides are based on the same essential chemical; some are diverse. Some products are approved for organic use and some are not.

To make matters even more complex each state has its own labeling process. Some states allow the use of certain pesticides and some don't. Various types of pesticides are registered for use on particular crops in some states and not in others. This information is available on the product label. In addition there will be specific information about

Spraying copper in our 1-acre (0.4-hectare) hop yard to combat downy mildew.

how to mix and apply the pesticide, how often to apply it, and the associated safety standards that must be adhered to. In many cases the person who applies the pesticide must be a certified applicator or under the supervision of someone who is. It is illegal to use a pesticide in any manner other than that stated on the label. Off-label use of pesticides is illegal and harmful to the environment and the applicator. It can result in a seizure of the crop, revoked applicator licensing, and substantial fines and penalties.

Fungicides and insecticides approved for organic use are available to combat many of the most pressing problems facing hop growers, but their numbers are limited, and the challenges of using many of them, such as phytotoxicity and killing beneficial insects, are often on par with the challenges posed by the synthetic chemicals approved for conventional use. To complicate matters, since hop production has been essentially nonexistent in the East in modern times, there are very few materials approved for use on hops in eastern states to choose from—whether you are farming organically or not—because the market has not been there to drive the research required to warrant the labeling. In addition, many of the diseases and pests that afflict hops are known for their ability to rapidly develop resistance to the materials used against them, requiring frequent rotation of types of sprays—difficult when you don't have many to choose from.

Whether you are relying on organic or synthetic chemicals, prevention is always the best choice since it will keep damage to the crop from occurring. But also, if it works, it will mean you have to spray less in the long run. And reaching for the most toxic thing available is not necessarily the best choice: you may end up killing the populations of beneficial insects that, combined with your preventive steps, will continue to keep pest populations under control. (See Chapter 9 for a thorough discussion of beneficial and nonbeneficial insects in the hop yard.)

You Are Not Alone

All this is overwhelming, especially for a new hop grower, but fortunately you are not alone. When trying to figure out whether or not you need to spray, what to spray, and when to spray it, consult with your local Extension agent. Even if your goal is to avoid synthetics at all costs or as much as possible (as Dieter and I have chosen to do), an Extension agent's regional wisdom can be invaluable. We have turned to ours many times.

Back in 1862 an act of the federal government called the Morrill Act granted control of federal land to states, allowing them to sell the land to raise money to endow institutions of higher learning on the condition that these schools would provide education on agriculture, science, and engineering. Each state has at least one land grant university. These universities operate a Cooperative Extension system, which is an educational network of local offices to provide practical, research-based information to farmers and other parts of the community, such as youth and consumers. Each land grant university has an agricultural program that generates research and guidelines for farmers, made available through its Extension offices. While the only Northeastern land-grant programs focusing on hops are UVM and Cornell, all Extension offices are equipped to help farmers understand which organic and conventional sprays are labeled for use on hops in a given state and help calculate spray applications. Extension agents are also extremely helpful when it comes to scouting for pests and disease, conducting and interpreting soil tests, and calculating soil amendments and fertilizer applications (for more information on university Extension offices offering information on growing hops, see the Resources).

Steve Miller, hop specialist, talks to a group of new hop growers in New York State.

CHAPTER 8

Disease

When you are putting in a new crop or expanding the plant diversity of your backyard, it can be daunting to read about all the potential diseases that will afflict the crop for which you have such high hopes. It may seem like you and your plants don't stand a chance. But while it is true that growing hops in the East is going to be challenging, especially if you are trying to avoid reliance on routine application of chemical sprays that may be harmful to the environment, it can be and is being done.

Traumatic or not, reading about the various diseases you may encounter is critical. And it's something you need to do before you plant, not after, because the main way to manage disease in your hop yard using sustainable practices will be through actions you take in the early stages. As discussed in previous chapters, planting the disease-resistant varieties that have been developed since hop production vanished from the East will be your first line of defense. Locating your hop yard to optimize drainage and exposure to sun and wind is also key, as is preparing the soil and suppressing the weeds prior to planting. Once the hops are growing, sharp eyes, quick action, and ongoing vigilant weed control will help you keep diseases in check when they arrive in your hop yard.

The major diseases that appear in hops in the East are either fungal or viral. A fungal disease is caused by a fungal parasitic organism that lives off a host. Some of the major diseases of hops are fungal in nature.

Downy Mildew

Downy mildew is the number-one enemy of hop farmers around the world, and the wetter it gets in the East, the better the conditions are for it to flourish. That's why agronomist Heather Darby, from UVM's Extension agency, focuses intently on the disease at the experimental hop yard she oversees in northern Vermont, on the Canadian border. We met up with her there one day to talk about disease prevention, and her efforts to test organic approaches. Darby walked through the hop yard, turning over leaves to search out signs of disease. "Everything bad in the hop yard first appears as little yellow spots," she exclaimed. "First you tell yourself it's just this or it's just that, then one day you see it full blown and say, 'No! It can't be!' But by then it is usually too late to stop it." She didn't have to look far. She soon came across a cluster of blighted leaves where downy mildew had struck.

Downy mildew is a sort of generic term for a group of funguslike parasites that specialize in moving through water to penetrate the surface of plants, then sucking the life out of them. In actuality, it is an oomycete, hailing from a group of microorganisms sometimes called water molds. So it is not exactly a fungus, but it acts like one, hence the name. Different types of downy mildew affect different species of plants. The scientific name for the type of downy mildew that goes after hops is *Pseudoperonospora humuli*. It is closely related to another type of downy

Heather Darby, agronomist for UVM's Extension agency, examines hop leaves at UVM's experimental organic hop yard at Borderview Research Farm in Alburgh, Vermont.

mildew, *Pseudoperonospora cubensis*, that afflicts farm crops such as cucumbers and squash. Although *P. humuli* and *P. cubensis* attacks are species specific, researchers are looking into whether it is possible for them to cross species.

Where *P. humuli* came from is anyone's guess, but it infects hop yards around the world, with the sole exception of Australia, New Zealand, and South Africa. It made its first appearance in Japan in 1905, then the following year popped up randomly on the other side of the world in Wisconsin. It hit England in 1920 and spread through Europe. In 1928 *P. humuli* arrived in New York, playing a major role in the ultimate demise of commercial hop farming in that state, and ten years later established itself in the hop-growing regions of the Pacific Northwest.

Downy mildew lives in hop rhizomes and spreads when rhizomes harvested from an infected hop yard are replanted somewhere else. The spores can also travel on the wind as well as on the clothes, shoes, and hands of people who handle infected plant material or visit an infected hop yard, then enter an uninfected hop yard.

Downy mildew first appears as a diseased shoot called a basal spike emerging from the ground.

Downy mildew advances from the roots into the plants, becoming systemic. Photographs by D. R. Smith, *Compendium of Hop Diseases and Pests*, American Phytopathological Society

Systemic downy mildew causes real damage when it reaches the cones, causing extreme browning and destroying the lupulin. Photographs by D. R. Smith, *Compendium of Hop Diseases and Pests*, American Phytopathological Society

What's the damage? A lot. Downy mildew stunts the growth of the hop bines, reducing the number of cones they can produce. It also impacts yield by damaging the cones themselves and in some cases can outright kill the entire plant. We had always grown single plants here and there on our farm and never had downy mildew. Then we put in our pilot hop yard and were free of downy mildew for the first two years. In the third year that yard was full of it and it began making a sporadic appearance in the 1-acre (0.4 hectare) yard. So we joined the ranks of hop farmers fighting downy mildew. You can and should try to stay free of downy mildew for as long as you can, but sooner or later you will get it, and from that point on it is a question of management and control. As a matter of fact, control of downy mildew is the principle on which many of the standard best management practices in the hop yard are based. To control downy mildew, first you must understand it. It is a diabolical pathogen that, like all other living creatures on this Earth, is working a number of different angles to perpetuate its existence. Its end goal is to reproduce itself as much as possible.

LIFE CYCLE

The disease is so hard to get rid of because it overwinters underground in the crown of the hop plant. Infected crowns have reddish-brown to black flecks and streaks in the white crown tissue next to the bark. The infection infiltrates developing buds, causing them to send up diseased shoots, called primary basal spikes. These spikes appear stunted, with yellowed leaves that are brittle and curl downward. Sporangia (spore sacs) develop on the undersides of these leaves as well as on the spike at night, when the temperature and moisture levels are right. Between midmorning and early afternoon, especially when it is wet and rainy, the sporangia release spores into the hop yard. The production of spores from the diseased spike and leaves continue throughout the conducive period, spreading the disease throughout the hop plant and between plants. These zoospores, essentially little swimmers with tails, really like it when there is standing water on hop leaves that are in close proximity to the basal spike producing the spores because they can swim through the water to enter the plant through one of the tiny openings in the plant's surface, called stomata—pores through which the hop breathes. This is how downy mildew spreads through the hop plant.

When secondary spikes (which look much like the primary spikes) emerge from the top of the plant or its side shoots, the infection is no longer on the surface of the plant but has gone systemic. As the disease progresses, it moves up through the plant. Aerial spikes develop on the lateral branches and trained bines in late spring or midsummer. Infected trained bines will stop growing and simply fall from the string they were previously climbing so vigorously. Lesions develop on infected leaves. Developing flower clusters turn brown and shrivel. If the infection didn't take hold until later in the season, only the bracts will turn brown and the cone will look striped. In severe cases, cones turn completely brown and harden. In some cases the plant dies of reduced energy reserves caused by disease.

Remember the disease triangle? When it comes to the pathogen known as downy mildew, the host is hops and the environment is wet and warmish. The question is how wet and how warm, and for how long. In general, if downy mildew is present in your hop yard the infection can spread significantly if conditions remain wet and humid at a moderate temperature of 65 to 76 degrees Fahrenheit (18.3 to 24.4 degrees Celsius) for four to eight hours. Sounds like spring in the Northeast, right? Right.

The first time I was reading all the stats about the conditions under which downy mildew spread, it was early June and it was raining, and warm. It rained and stayed warm all week, day and night. When it finally stopped raining I went out on the deck and checked the rain gauge. We had gotten 3 inches (7.6 centimeters) of rain in four days. In our small hop yard, downy mildew had taken over and had started to appear in the larger yard, too. We didn't know where it came from, but it had undoubtedly arrived. I was really starting to understand why hop farmers are so neurotic about water. Hops are really thirsty plants, but they do not want to go swimming. What's a hop farmer to do?

I thought about the disease triangle some more. We could not eliminate the host. The host is our crop. Some of the varieties we had planted were tolerant of downy mildew and some were susceptible. We couldn't exactly pull the susceptible ones up—at least not yet. And unfortunately we cannot change the environment (although we seem to have already changed it for the worse). The only thing we can control at this point is the pathogen, so that is what we set out to do.

PREVENTION AND CONTROL

The first step in controlling downy mildew is to try to keep it out of your hop yard in the first place. If you've acquired disease-free rhizomes through the National Clean Plant Network, that's a good first step, but don't forget to think beyond your crops. Downy mildew can infect wild hops growing near your hop yard. Monitor these wild hops as closely as you monitor your own hop yard. If there are signs of infection it is a good idea to eliminate the wild hops, if possible.

Although you can't prevent downy mildew spores from blowing in on the wind, you can avoid introducing them into your hop yard yourself. If you visit other hop yards or go out to collect wild hop cones for a home brew, change your clothes and shoes and wash your hands and tools before you enter your own hop yard. Think twice about who enters your hop yard and if they might inadvertently be carriers of downy mildew spores. Although hop yards are relatively far and few between in the eastern United States, it doesn't hurt to check with people entering your hop yard to make sure they have not just come from another hop yard. Extension workers aiding with activities such as soil testing and pest scouting should know about precautions against the spread of disease, but field workers or contractors installing trellising may not be aware of the issue.

Also think twice about what varieties you plant, choosing hop varieties that are bred to have a natural resistance to downy mildew. No hop is completely resistant, but some varieties are much more susceptible than others. We can only hope that brewers in the East will care enough about brewing beer made with local ingredients that they will develop recipes that make good beer out of varieties that the region's farmers can successfully grow.

You may take all these precautions and still end up with downy mildew—a reality that is driving downy mildew research in Oregon and Europe, where breeding varieties of hops that are tolerant of and resistant to downy mildew has been a huge priority for researchers for quite some time. Heather Darby at UVM is researching organic means of controlling downy mildew. In addition to choosing resistant

TABLE 8.1. Downy Mildew Resistant Hop Varieties

Variety	Resistance Level	Type
Brewer's Gold	moderately resistant	bittering
Cascade	moderately resistant	aroma
Chinook	moderately resistant	dual purpose
Columbia	moderately resistant	aroma
Fuggle	resistant	aroma
Hallertauer Gold	resistant	aroma
Hallertauer Magnum	resistant	bittering
Hallertauer Traditional	resistant	aroma
Liberty	moderately resistant	aroma
Newport	resistant	bittering
Opal	resistant	dual purpose
Perle	resistant	dual purpose
Spalt	resistant	aroma
Willamette	moderately resistant	aroma

varieties, there are management strategies that can help control downy mildew in the hop yard.

In fact, many cultivation practices standard in the hop yard have their roots in downy mildew control, including the spring activities of crowning and pruning. The most important thing you can do to control downy mildew is scout your hop yard frequently, keeping a vigilant lookout for primary basal spikes and, as the season advances, the secondary spikes that may appear higher up on the plant. Wherever they appear they must be cut down, removed from the hop yard, and destroyed. Do not compost downy mildew–afflicted vegetation; it is too risky. If the temperature of your compost does not get high enough and you put the compost in the hop yard, you will recontaminate your hops. The best way to destroy it is to burn it.

If you know heading into the season that you already have downy mildew lurking belowground, you can get a step ahead of it by crowning the plant. When crowning, a mechanical spinning blade attached to a motor in an arrangement similar to a weed trimmer removes the top ¾ to 2 inches (1.9 to 5.1 centimeters) of the crown. This intense underground pruning process helps prevent the hop from sending up an early crop of diseased shoots. This can be accomplished by hand with a garden knife. Larger hop yards crown their plants routinely to remove the previous year's growth as well as cutting back early shoots, which are not generally the strongest.

In addition to controlling downy mildew belowground, in the plant's root base, there is also much you can do at ground level—including pruning any diseased shoots that emerge from the ground and keeping the base of the bine weed free. Allowing weeds to grow around the base of the bine creates a moist, shaded environment perfect for harboring the downy mildew spores that are making their way up from ground level. Pulling all weeds allows sun in and improves air circulation, keeping the area dry and unfavorable to the spread of downy mildew. In addition, once you have selected and trained your bines of choice onto the string, it is important to remove all subsequent shoots the hop sends up, again to thwart any buildup of vegetation on the ground.

Once the trained bines have reached a height of 4 feet (1.2 meters) or so, start to strip the leaves off the base of the hop bine itself. As the bine grows higher, continue to strip leaves—working your way up until the bine is without leaves from the ground to a height of about 4 feet. Like removing vegetation on the ground, this process keeps downy mildew from progressing from the ground up into the plant. This is a lot of work, but it can all be done by hand in smaller hop yards. Most large hop yards use chemicals to kill the weeds and even to strip the leaves from the hops themselves.

Hop farmers have also experimented with bringing sheep into hop yards to eat the lower leaves from the hop bines and the weeds at the base. The long period of strict confinement it would take to get sheep to eat weeds down to bare dirt would probably put the hop bine at risk of being eaten as well, but sheep will do a pretty good job of eating the bottom leaves off the hop plant and shouldn't cause much damage as long as there is plenty of vegetation on the ground for them, too. The advantage of sheep over goats is that sheep don't tend to stand on their hind legs to browse what they cannot reach with four feet on the ground, which limits the height up to which they will eat the hop vegetation. That being said, you can't totally count on sheep to stick to lower heights: I have certainly seen sheep on their hind legs eating out-of-reach apples from trees on our farm. Another major concern when using sheep in the hop yard is that copper, a common spray used to control downy mildew, is highly toxic to sheep. Once the decision is made to spray copper in the hop yard, sheep must be removed from the hop yard for the remainder of the season.

Once the vegetation is cleaned out, put down a layer of mulch and rake it into a hill around the base of each plant. This will discourage the emergence of any further spikes. Although keeping the base of the hop plants clean of vegetation helps prevent disease, this practice also can have a negative impact because it deprives beneficial insects of a good environment to live in while they do their work of preying on insect pests in the hop yard. As with everything else in farming, the trick is to achieve a balance. Assess what your greatest risk is and do what you have to do to minimize it.

Another strategy for controlling downy mildew is to control moisture levels. Although it is clearly impossible to control rain and humidity, you can choose how you water your hops. Because of the threat of downy mildew and its preference for standing water on leaves, hop farmers avoid overhead, sprinkler-type irrigation. Drip irrigation that runs water directly onto the ground but can be lifted for crowning is the standard in many hop yards (see Chapter 5 for more on irrigation design).

FUNGICIDES

If you think you are looking at a bad outbreak of downy mildew, you'll need to keep up with the cultivation practices outlined above, but it also may become necessary to spray a fungicide. There are not a wide range of fungicides permitted for use on hops in the East, and most only work when applied preventively, before the disease strikes or right after it makes its first appearance. These nonsystemic fungicides treat the surface of the plant and work best at preventing downy mildew from taking hold before it gets bad. Others, known as systemic fungicides, are better for killing downy mildew once it gets into the plant's system; there are far fewer of these. Only a few types of fungicides are considered organic. The primary fungicide used in organic hop yards is copper.

The benefit of fungicide for downy mildew is clear. It kills the arch enemy of hops. However, fungicides have distinct disadvantages as well. One of these disadvantages is that fungicides also kill beneficial insects. You can spray fungicide and beat back the downy mildew but only end up jumping from the frying pan into the fire when the weather changes and a new pest comes to town. The health of your beneficial insects as well as human, environmental, and economic issues are more good reasons to choose to plant hop varieties bred to tolerate downy mildew. The more tolerant your hop varieties are, the less likely they are to contract a severe case of downy mildew and, if they do, the less you will have to spray them to bring the infection under control.

There are a couple of important things to remember if you are going to spray fungicide. Since fungicides tend to be most effective on downy mildew in hops if they are sprayed as a preventive, you'll need to monitor the weather. If you know downy mildew is present in your hop yard and you see a stretch of weather on the horizon that is going to be conducive to downy mildew reproduction, you should spray fungicide before that weather hits. The other really important thing to remember is that downy mildew is quick to develop resistance to fungicides, some more so than others. If you just keep on spraying the same fungicide over and over, it won't be long before it has no effect. You have to keep rotating fungicides.

Remember, spraying pesticides—whether fungicides, insecticides, or herbicides—is complicated and can be dangerous. There are plenty of people out there trained to help you with this, from the manufacturers to Cooperative Extension agents. Don't wing it.

Powdery Mildew

Powdery mildew is a major problem for hop growers worldwide. It tends to afflict areas that are drier; therefore, it is less of a problem in the humid eastern region. As of this writing it has not yet impacted fledgling twenty-first century hop growers in the East. The disease is a challenge to fruit and

True to its name, powdery mildew looks like talcum powder on the surface of the leaf, first appearing as white spots, then covering the entire leaf. Photograph by W. F. Mahaffee, *Compendium of Hop Diseases and Pests*, American Phytopathological Society

Once established, powdery mildew will move up the bine, covering the entire plant. Photographs by D. H. Gent, *Compendium of Hop Diseases and Pests*, American Phytopathological Society

vegetable growers in the region, but it is species specific. It used to be thought that the form of powdery mildew that affects hops went by the scientific name of *Podosphaera macularis* and also infected chamomile, strawberries, and hemp, all of which grow wild and are cultivated in the East. But new research has shown that *Podosphaera macularis* only affects hops and cannabis. Some research indicates that powdery mildew afflicting the genus *Cucurbita*—which includes crops such as summer and winter squash, pumpkins, and cucumbers—can jump to hops and has done so in North Carolina, Michigan, and Washington. This type of powdery mildew is called *Podosphaera fusca*. This is not good news for our farm since pumpkins and winter squash are grown in great numbers on the apple farm our land is carved from. Late in the season the great, broad leaves on the pumpkin and squash plants often appear grayish white and blasted as powdery mildew takes its toll.

Despite strict quarantines, powdery mildew finally made its appearance in greenhouses in Washington in 1996. This prolific fungus, which causes extreme reductions in crop yield and quality, was spotted in a Yakima Valley hop yard in June of 1997 and by the end of July had rampaged through every hop yard in the state, destroying $10 million worth of hops on 2,000 acres (810 hectares). The following year it took over in Oregon and Idaho. Needless to say, the hop industry in the Northwest is very focused on fighting powdery mildew. In fact, Paul Matthews's lab at Hopsteiner, in the City of Yakima's industrial zone, has a disease challenge greenhouse, where he and his staff put hop seedlings on trial by introducing "a hop plant furry with powdery mildew" into the greenhouse. "We throw out the seedlings that get powdery mildew," explains Matthews. "The ones that don't get cleaned up and go out to the field. We do this at the greenhouse here in town because no grower wants a bunch of mildew-induced hops anywhere near his hop yard. It just feels bad."

Will powdery mildew on hops make its way to the East? Is it already here living on other plants and the occasional wild hop? Will it jump from our pumpkins into our hop yard? The answer to at least the first, if not all, of the above questions is likely yes. So what do we do?

Powdery mildew in the hop yard can (to a certain extent) be prevented, at least for a while, and it can definitely be controlled; however, it cannot be eliminated. So, much like downy mildew, once it turns up in the hop yard, it becomes a management issue, and to manage it you need to understand it.

Life Cycle

Like downy mildew, powdery mildew thrives in a moist environment; unlike downy mildew it can also develop and spread during spells of dry weather because it does not need direct contact with water to reproduce or infiltrate the plant. Between 64 and 70 degrees Fahrenheit (17.8 and 21.1 degrees Celsius) is powdery mildew's ideal temperature, but it can reproduce when the temperature is in the 46 to 82 degree Fahrenheit (7.8 to 27.8 degrees Celsius) range.

Its fungal spores, called conidia, blow into a hop yard on the wind and overwinter underground on the buds on the rhizomes, in the soil, and on plant litter on the ground around the hills. In the early spring it makes its appearance on emerging shoots. Infected shoots, called flag shoots, are stunted, pale yellow, and look as if they have been dusted with white talc. This white powdery substance comprises the fungus's spore-producing masses. When the conditions are right, spores are released and spread through the hop yard—again by the wind but also by water splash and people and equipment moving through the hop yard. Conditions most conducive to spore release are cloud cover, too much moisture in the soil, an overabundance of nitrogen, dense vegetation, and temperatures between 64 and 70 degrees Fahrenheit. Powdery mildew particularly likes it when the daytime and nighttime temperatures are similar and

As with downy mildew, if left unchecked, powdery mildew will ultimately destroy the hop cone. Photographs by W. F. Mahaffee, *Compendium of Hop Diseases and Pests*, American Phytopathological Society

range between 50 and 68 degrees Fahrenheit (10 and 20 degrees Celsius). Powdery mildew can be hard to detect because the flag shoots are only produced intermittently and may be quickly overgrown by other emerging hop shoots as well as weeds.

Later in the season powdery mildew appears as small, pale, yellowish spots on the tops of the leaves. This discoloration, referred to in plant science lingo as chlorotic, indicates that the plant is lacking in nutrients. Although this can sometimes indicate a problem with the soil, in the case of a fungus such as powdery mildew, which is essentially a parasite, it means that something else is eating the nutrients that should be going to the plant. Without those nutrients, the plant is unable to produce sufficient chlorophyll, which is what makes it green—hence the pale spots on the leaves. The lack of chlorophyll also reduces the plant's ability to feed itself with the carbohydrates it produces through photosynthesizing sunlight. Over time this slows the plant's growth and kills off leaves.

As the powdery mildew takes hold, it will actually become visible on the leaves as powdery white spots, similar to the white dusting on the flag shoot. These white spots will first appear on the undersides of leaves but will also become visible on the tops of leaves as well as on other parts of the plant, including stems, buds, and flowers. These white powdery spots produce spores. Over time the spots get bigger and more spots appear on other leaves, producing more spores, and so it goes.

Powdery mildew is unlikely to kill the plant itself, but it significantly damages the crop by reducing both the yield and the quality of the cones. If the infection gets into the burrs or the young cones, it will either kill off the cone or deform it. Sometimes the white powdery fungal growth will be visible on the cone, but sometimes it is only visible under the bracts and bracteoles and sometimes only with magnification. Damaged cones can appear reddish-brown, but sometimes they look normal at harvest, then turn brown after kilning.

PREVENTION AND CONTROL

As with downy mildew the first line of defense is to plant disease-resistant varieties. Wouldn't it be great if the same varieties that are resistant to downy mildew were resistant to powdery mildew? Yes, it would, but that is not the case. Cascade, Liberty, Newport, and Nugget are tolerant. Note that Cascade is the only variety tolerant of downy mildew that is also tolerant of powdery mildew; hence, its popularity with growers. Fortunately it is also popular with brewers because it is dual purpose, with bittering qualities as well as a big hoppy aroma.

You should also be careful not to overdo it with the nitrogen. Even though hops need a huge amount of nitrogen, there is such a thing as too much. If you overload the hop yard, the plant growth will be spongy and succulent, which is conducive to the development of powdery mildew.

Since there are not yet very many hop yards in the East, and no current powdery mildew outbreak in those that do exist, you may think that means you are safe from powdery mildew. But remember, there are wild hops growing in the region, some of which may be native, but most of which escaped from cultivated hop yards during the 1800s. Most hop growers have a soft spot in their hearts for wild hops. But remember, if there are wild hops growing in the vicinity of your hop yard and they become infected with powdery mildew, they will spread it to your hop plants. So once again, some growers recommend eradicating any wild hops growing near your hop yard. If you can't bear to do this, at least monitor the wild hops carefully so that if there is an outbreak of powdery mildew or something else, you are aware of it. Remember, most feral hops have the same genetics as Cluster, which is the variety many colonists brought with them from Europe, and Cluster is notoriously susceptible to both downy and powdery mildew.

Sunshine and air circulation are enemies of powdery mildew. Although spores can arrive on the wind, good exposure to wind and sunlight will keep

Powdery Mildew Home Remedies

Powdery mildew is a fungus that afflicts numerous garden plants. As a result the gardening community has come up with a number of home remedies to treat it. Currently there is no research into whether or not these remedies work on hops, but if you only have a few hop plants it might be worth experimenting.

MILK

Although the exact reasons for why it is effective are still being debated, it has been proven that spraying milk diluted in water on afflicted vegetation controls powdery mildew. The same effect has also been attributed to whey. The best ratio of milk to water for fighting powdery mildew is not yet known, but people have had success with ratios as low as 10 percent milk to 90 percent water and as high as 40 percent milk to 60 percent water. The main thing is to spray the milk onto the plant in full sun, as it seems that the key to success is the interaction of compounds within the milk and the ultraviolet rays of the sun.

BAKING SODA

Baking soda, sodium bicarbonate, is also effective against powdery mildew. Combine 1 tablespoon (15 milliliters) of baking soda with 1 gallon (3.8 liters) of water and mix with 2½ tablespoons (37 milliliters) of horticultural oil. Put this solution in the spray attachment on your hose and spray. Test a small area a few days prior to spraying an entire hop plant to make sure the mixture does not cause phytotoxicity.

To make horticultural oil, mix 2½ to 3 tablespoons (37 to 45 milliliters) of seed oil (such as canola, cottonseed, or soybean) with 1 gallon of water and ¼ teaspoon (1 milliliter) of liquid soap, such as an all-natural Castile soap, which is vegetable-oil based.

them from taking hold on the plant's surface. When you lay out your hop yard, don't skimp on space in between plants. If, as they grow, the plants become crowded together, the thick vegetation will limit air circulation and sunlight exposure, and powdery mildew will proliferate.

For the same reasons it is also important to remove vegetation from the base of the plant, including both weeds and hop shoots that you are not training. Not only will this increase sunlight and air circulation, it will make it easier for you to spot those diseased flag shoots when they come up. These flag shoots must be removed immediately, taken from the hop yard, and destroyed. Understand that once you see a flag shoot or white patches on the leaves of the plant it is too late. You have powdery mildew in your hop yard. Your purpose at this point will not be to eliminate it but to minimize it. Once it occurs, removing the plant growth on the ground at the base of the trained bines keeps the powdery mildew from spreading upward. As with downy mildew, it also makes sense, once the plant has reached a height of at least 6 feet (1.8 meters), to begin stripping the leaves off the bottom of the bine to a height of about 4 feet (1.2 meters) off the ground. Keeping the area on the ground at the base of the bine free of vegetation also ensures that if you are forced to spray a fungicide the area will be open and the spray will be able to reach the hop plant itself.

FUNGICIDES

All commercial hop yards infected with powdery mildew find it necessary to combine cultivation practices such as those described above with a fungicide spray program. As with downy mildew the most effective spray program is a preventive one. If you know you have powdery mildew, start your spray program as soon as the shoots emerge—then watch the weather. When weather conditions conducive to the spread of powdery mildew begin to line up, it is time to spray. What you choose to spray will be decided by several factors. The first factor will be whether you are committed to organic practices or are using conventional practices. Even if you are using conventional practices you may want to start with organic fungicides. Most organic fungicides are best suited to preventing the spread of powdery mildew rather than killing it, and they can be less harmful to the environment. You will then need to consider what fungicides are legal in your state and which of these is labeled for use on hops. As with downy mildew, powdery mildew is quick to develop resistance to fungicides so it is important to rotate the fungicides you use. Not to do so is irresponsible. By fostering fungicide resistance in a pathogen that destroys a commercial crop, you are not only harming yourself but also harming others who are trying to make their living growing hops.

Verticillium Wilt

Verticillium wilt is yet another disease of hops caused by a fungus. This disease affects a broad range of cultivated and wild plants worldwide ranging from alfalfa to watermelon. There are two species of the fungi: *Verticillium albo-atrum* and *Verticillium dahliae*. The symptoms of the disease range from mild to severe depending on the virulence of the strain and the susceptibility of the plant. It normally strikes fairly late in the season when the cones are in development, causing yellowing and wilting of the plant moving from the ground up. In severe cases the plant will die. The disease can be dormant in the soil or be brought into the hop yard from new infected rootstock or plants as well as by people who have been in a contaminated hop yard.

LIFE CYCLE

Some types of fungi can live in the soil in what is referred to as a "survival structure" for several years. The verticillium wilt mode of survival structure is mycelium, a vegetative part of the fungus that stays dormant in the soil waiting for the root of a potential host plant, such as the hop, to come meandering by. *Verticillium albo-atrum* can lie dormant in the soil for four years. *Verticillium dahliae* can hang around in its survival structure for as long as fifteen years. Sensing the proximity of the root, the fungus comes to life and enters it, usually through a minor wound the root sustained while digging through stony dirt. It travels through the root into the plant.

The first visible symptoms generally come on at first flowering but can occur when the plant is

Verticillium wilt appears as severely wilted, curled leaves that quickly die. Photograph by S. Radisek, *Compendium of Hop Diseases and Pests*, American Phytopathological Society

stressed. The first thing you will see is the yellowing of the plant's lower leaves. Once the fungus infiltrates the plant's vascular system, it interferes with and in severe cases eventually blocks off the circulation of water and nutrients through the plant tissue, killing the plant. The yellowing starts on the sections of the leaves that lie between the leaf veins and gives the leaves a streaked appearance. Eventually the edges of the leaves begin to curl upward. The bine itself becomes swollen, and if you cut into it you will see that the interior of the bine is streaked with brown. Eventually the leaves become entirely brown and fall from the bine. The cones themselves literally wither on the bine. When the plant dies, the fungus, presumably well nourished on all the resources that should have been going into the plant, retreats back into its survival structure, where it waits for its next victim.

PREVENTION AND CONTROL

If you believe you have verticillium wilt in your soil, the best option is to avoid planting hops there; choose another location for your crop. There is nothing that can be sprayed on hop plants to kill verticillium wilt once it has gotten into a plant. Soil fumigation will kill it in the soil, but soil fumigation uses highly toxic chemicals that are not only bad for people and the environment but are bad for the soil itself.

One very sustainable means of controlling verticillium wilt is crop rotation. By planting crops that are not potential hosts for verticillium wilt over a period of years you can use the land productively

A hop bine suffering from verticillium wilt appears swollen. Photograph by S. Radisek, *Compendium of Hop Diseases and Pests*, American Phytopathological Society

Once it strikes a hop yard, verticillium wilt is very difficult to eradicate. The best recourse is to relocate the hop yard. Photograph by S. Radisek, *Compendium of Hop Diseases and Pests*, American Phytopathological Society

while depriving verticillium wilt of a host until you can be sure it has died out. But hops are perennial plants and do not fit into a crop rotation model. You'd have to rotate other nonsusceptible crops for more than four or fifteen years, depending on the verticillium species, before planting hops.

As a safeguard against the eventual appearance of verticillium wilt in the hop yard, the best and really only option you have is to plant tolerant varieties of hops. Tolerant varieties include Cascade and Perle. Fuggle is particularly susceptible.

The fungus spreads when roots from an infected plant touch the roots of a neighboring uninfected plant. It can also move into different locations in the hop yard through the cultivation of the soil. In some cases, its spores travel on the wind. In addition, common weeds such as pigweed and lamb's-quarter can contract verticillium wilt and spread it through the hop yard. To make matters even trickier some plants, including hops and weeds, can have verticillium wilt and not show it. They can, however, spread it. So if you think you have verticillium wilt in the hop yard it is best to remove weeds and hop vegetation left behind after harvest from the hop yard.

Hops pull a lot of nitrogen from the soil, and because much of this nitrogen remains in the vegetation of the plant after harvest, hop farmers often return the vegetation to the field and let it decompose. This is not a good idea if you have verticillium wilt in your hop yard as it will just reintroduce the pathogen into the soil. Research has shown that small amounts of verticillium wilt spores can even survive composting, so returning even composted hop vegetation that has been exposed to verticillium wilt to the yard should be avoided.

Viruses

A virus is a microscopic particle that only becomes active when inside its host of choice. Hop viruses only come to life and begin to reproduce when they are present in the plant itself.

APPLE MOSAIC VIRUS

Apple mosaic virus is the most significant viral disease impacting hops around the world. It also affects apples, pears, and roses but does not naturally transmit from one species to another. Plant propagation is the main way this disease spreads, and the only way to make sure you'll keep it out of your hop yard is by purchasing only certified healthy rhizomes and plants. This is worth doing because if unleashed, the destructive virus is capable of reducing your yield of cones as well as the production of valuable alpha acids by as much as 50 percent. Sometimes plants can be infected with apple mosaic and hop mosaic viruses at the same time, and the loss becomes even more severe. Its primary symptom is a yellowing of leaves in an oak-leaf-like pattern. It becomes most severe when the temperature remains under 80 degrees Fahrenheit (26.7 degrees Celsius) for a period of time, then becomes hotter. Insects do not spread this disease through the hop yard—but people do on their hands, tools, and equipment.

CARLAVIRUS COMPLEX

The carlavirus complex is a family of viruses that includes hop mosaic virus, hop latent virus, and American hop latent virus. Carlavirus is short for "carnation latent virus." A latent virus is one that has the ability to lie dormant within a plant for a period of time, then suddenly begin to reproduce at a high rate. When the virus is dormant the host does not show any symptoms but can still pass the virus onto a new host. This is a great trick. The three hop viruses included in the carlavirus complex can occur in the hop yard in various combinations. The hop latent and hop mosaic viruses are found worldwide, whereas the American hop latent virus is only found in North America.

The most common way the carlaviruses enter the hop yard is through infected plants and rhizomes. Once in the hop yard the virus can be spread by aphids, so control of aphids can reduce the virus's

Apple mosaic virus first appears as a yellowing of the leaves and can ultimately cut the yield of alpha acids in half.
Photograph by D. H. Ghent, *Compendium of Hop Diseases and Pests*, American Phytopathological Society

Symptoms of the carlavirus complex include yellow spots on the leaves and weak growth. Photograph courtesy of David Gent, USDA Agricultural Research Service, Bugwood.org

spread through the yard. The viruses are also spread through mechanical cultivation. While hop latent virus and American hop latent virus do not cause obvious symptoms, hop mosaic virus causes a pattern of yellow mottling on the leaves and weakened plants with reduced bine growth. In some cases the bines will not even have the strength to wind around the string.

Golding, and varieties descended from it, seem to be most susceptible to hop mosaic virus. Infection with the carlaviruses slows plant growth, making it hard to establish new plantings, and reduces yield. As with all viruses, the best way to protect your hop yard is to only obtain new plant material from certified healthy plant sources. If you find a plant in your hop yard stricken with one of these viruses the best thing to do is to remove the plant as well as any neighboring plants. Since the disease can be transmitted from plant to plant by mechanical activities such as pruning, it is important to clean tools (and hands or gloves) in between plants.

HOP STUNT VIROID

Hop stunt viroid is a serious disease that afflicted hops in Japan and South Korea before being discovered in North America in 2004. It stunts the plant as well as the cones, dramatically reducing the alpha acid. Like the carlaviruses, the hop stunt viroid is a latent viroid. A hop plant can be infected with the viroid, and spread it to other plants, for several years before it shows any symptoms. Although the plant may appear healthy it will produce less alpha acids, even before other symptoms strike. Once it shows symptoms, the plant itself will appear stunted, with curling, yellowed leaves, and produce small cones. The bine often has trouble climbing because it does not develop trichomes. The viroid is present in the sap of the plant and is spread through the hop yard as workers prune through the sap that gets on the tools. The viroid usually enters the hop yard on diseased plants and rhizomes. If the hop stunt viroid shows up, the best thing to do is to remove the diseased plants. Unfortunately, already infected plants could still appear healthy while continuing to spread the disease.

It may seem like everything is out to get your hops, but remember, the best defense against disease is a well-nourished, carefully tended plant grown from healthy stock. Maintaining a healthy soil ecology rich in organic matter and populated by beneficial microbes will go a long way toward fending off disease. Scouting your hop yard regularly and keeping a sharp lookout for the telltale signs of disease will enable you to act quickly to control an outbreak and keep it in check.

A hop plant can be infected with hop stunt viroid without showing symptoms and still be contagious for several years. When symptoms do start to show, the plant will exhibit curling, yellowed leaves and produce small cones. Photograph courtesy of David Gent, USDA Agricultural Research Service, Bugwood.org

CHAPTER 9

Insects

A hop yard has its own ecosystem. Yes, the hop bines constitute a monoculture, but unlike a crop that is planted annually, the hop bines stay in place over time, and the rows in the hop yard have a vegetative ground cover. Depending on weather conditions the hop yard attracts a variety of insect populations. Some of these insects are there to feed on the hops, and other types are there to eat those that eat hops. The trick is to control the damaging insects while nurturing the beneficial insects. This balancing act has to be performed throughout the growing season as conditions change, insects proceed through their life cycles, and weather patterns ebb and flow. Coming down too hard with a pesticide on one insect that happens to be causing a lot of damage at the time could kill off your population of beneficial insects, leaving your hop yard vulnerable to a secondary outbreak of the original pest or allowing an outbreak of a different kind of insect pest taking advantage of the opportunity posed by both a plentiful source of food and a lack of predators.

Different insects make themselves known at different stages of the growing season, and varying weather conditions favor the ascendance of one type or another. Depending on the weather and what went on in the previous season, each growing year presents its own set of challenges. Controlling insects in the hop yard is a big job that requires a lot of background knowledge combined with constant scouting for insects in the hop yard while keeping a continuous eye on the weather. As with diseases, there are a few major players you need to become very familiar with.

Hop Aphids

The hop aphid's Latin name is *Phorodon humuli*, but it is commonly called the damson hop aphid, named after the plum. How does an aphid come to be named after a plum? Somehow, during the course of aphid evolution and adaptation, this particular species of aphid came to specialize in eating hops during the summer while surviving the winter by hibernating in nearby plum trees—as well as other species of *Prunus*, a genus of trees and shrubs that includes producers of stone fruit such as cherry, peach, nectarine, and apricot. This curious niche seems to be working out very well for the hop aphid because it has been plaguing hop yards in the Northern Hemisphere for well over one hundred years. Nineteenth-century hop growers referred to the green, translucent bugs as hop lice and fought them with tobacco smoke and kerosene. In the 1950s hop growers in the Northwest pulled out the big guns and started spraying them with organophosphates and the like, but the aphids, their secret weapon being exponential reproduction on a staggering scale, were able to nimbly develop a resistance to just about every toxic chemical thrown at them.

Why go to such extremes to kill this tiny green bug? Because hop aphids are bad news in the hop yard. They are known to spread the carlaviruses, but

Hop aphids feed by sucking sap from the hop plant. They do not metabolize all that they eat. The excess, secreted as waste, is called honeydew. Photograph by David Fenwick

that is minor in comparison to the damage they cause by feeding on the hops. Hop aphids literally suck the lifeblood (sap in plant lingo) out of the hop plant, depriving it of the nutrients it needs to grow. To add insult to injury the hop aphids then excrete a sticky slime, generously referred to as honeydew, that covers the surface of the plant's vegetation. This coating can interfere with photosynthesis. In some cases the honeydew itself can pose an even more serious threat. According to Gayle Goschie, an outbreak of hop aphids on Goschie Farms left such a thick layer of honeydew on the hop vegetation that, when heated by the sun, it "literally fried the plants." But it gets worse. Honeydew provides an excellent medium for the growth of another dreaded affliction of hops—sooty mold. Sooty mold is a black fungus that looks like soot, hence the name. It does not actually harm the plant, although a good covering of it on the leaves, as with the honeydew, interferes with photosynthesis. But the real problem comes when the sooty mold follows the aphids' honeydew trail into the hop flowers themselves. The cones will turn brown, and their interiors, where the precious lupulin collects, will become riddled with black mold.

LIFE CYCLE

Hop aphids have a complex and highly effective annual schedule. They overwinter on *Prunus* as eggs. In early spring wingless female hop aphids hatch and feed on *Prunus* leaves. The wingless female aphid is capable of asexually producing male and female aphids with or without wings depending on what feeding and migration stage the population is at. In the spring the wingless females reproduce asexually and give birth to several more generations of wingless females who also begin to asexually reproduce, boosting numbers. In preparation for migration from *Prunus* to hops the aphids begin to produce females capable of flying to the nearby hop yard to feed for the summer months. When the temperature is right these winged females migrate to the hop yard. Some species of aphids are known for migrating great

A black fungus called sooty mold feeds on the honeydew left behind by hop aphids. If a hop aphid infestation takes place after the cones are developed, sooty mold can infect the cones, destroying their value to brewers. Photograph by W. F. Mahaffee, *Compendium of Hop Diseases and Pests*, American Phytopathological Society

distances on the wind, but hop aphids only travel about ½ mile (0.8 kilometer) or less. Once in the hop yard the winged females, who are also asexual, start producing wingless asexual females (no point in investing energy in producing wings when you are not going anywhere for a while). This is why most of the aphids found in hop yards during the summer are wingless. These wingless aphids continue to reproduce more of the same, and so it goes. Needless to say, the population builds quickly.

The aphid feeds by sticking its piercing, sucking mouthpart, called a stylet, through the surface of the plant's leaves, stems, and cones and using it like a straw to suck up the plant's sap. Both the nymphs, which are the baby aphids, and the adults feed on the plant. The aphids eat more sap than they can process, and the excess comes out the other end as the sugary solution called honeydew. If an aphid infestation is severe, this coating of honeydew can actually give the hop vegetation a shiny appearance. As they sense the summer coming to an end, the aphids start giving birth to both winged females and winged males. Everyone with wings flies back to *Prunus*. In the safety of the branches of their winter home the winged male and female aphids mate. The female aphids lay eggs that will hatch the following spring. Wingless female aphids will emerge and asexually reproduce, once again beginning to produce winged females as the time for migration approaches. And so the cycle continues.

Prevention and Control

There are essentially no varieties of hops that can tolerate hop aphids, which are controlled in the hop yard by two conflicting activities: cultivating populations of beneficial insects that prey on aphids and spraying insecticides (organic options are available).

Commercial hop growers put up with aphids in the hop yard during the growing season to a point, but once flower formation begins, a zero tolerance policy kicks in. The risk of damage to the cones from sooty mold is simply too great. It is critical that aphids in the hop yard be brought under control before the flowers form because once the aphids get inside the hop cones it is harder for predatory insects, which are often larger than the aphids, to get at them to eat them. It is also impossible, at this point, to kill them with insecticides, which will not penetrate the surface of the cone.

It is important to make your hop yard as attractive as possible for beneficial insects. These insects appreciate having vegetation on the floor of the hop yard. Unfortunately, the things that beneficial insects like are also desirable for other life forms that harm hops, such as downy mildew. The trick is to find the right balance. If you eliminate all the vegetation from the floor of the hop yard in your efforts to control the spread of downy mildew, you will inhibit your population of beneficial insects and may increase your chances of suffering damage from insect pests that feed on hops.

Bugs that prey on aphids will not be attracted to your hop yard until there is a sufficient number of prey present in the yard to attract their interest, so the aphids are always going to have a head start on the predators. Even if you have a great population of beneficial insects, aphids reproduce so fast their predators may not be able to keep their numbers down to a level that is safe for your hop cones. In that case you will need to spray. Predators of aphids include the aphid midge, ladybird beetle (transverse, convergent, multicolored Asian, and seven-spot), green lacewing, brown lacewing, hoverfly larvae, and parasitoids such as the aphid wasp. There are also what entomologists refer to as "predatory true bugs"—such as the minute pirate bug, the big-eyed bug, and the damsel bug—as well as the predatory fly, spiders, and other arthropods.

It is possible to lure beneficial insects to the hop yard by including the types of plants they are attracted to in the row cover vegetation in your hop yard. For example, yarrow attracts ladybird beetles as well as parasitic wasps and numerous other beneficial insects. Caraway and marigolds attract predatory true bugs. It is possible to purchase beneficial insects and release them in the hop yard, but timing the release correctly can be difficult. If the prey species is not yet high enough to sustain them, the predatory insects you release will simply fly away in search of food.

To find out if you have aphids in your yard, be on the lookout. Get a jump on things by putting out insect traps in the spring, once the temperature climbs above 58 degrees Fahrenheit (14.4 degrees Celsius), so that you can monitor the population level as they arrive in the hop yard. Hop aphids prefer cool, damp weather, so they are more likely to be a problem under these conditions. Scout for aphids weekly, checking the undersides of leaves. Look for small, soft-bodied, pear-shaped insects that range in color from pale to dark translucent green with dark green stripes running lengthwise on the abdomen. The nymphs are even tinier and appear white. The winged hop aphid is darker in color.

Commercial growers in the Northwest rely heavily on IPM to control hop aphids. One of the central tenets of IPM is to determine at what level the population of a type of pest will cause sufficient economic damage to warrant spraying. After much research the growers in the Northwest have determined that magical number to be the presence of eight to ten aphids per leaf. This is called the "economic threshold." It essentially means that the costs involved in spraying, which are significant, will be more than covered by the value of the crop you will save. Put another way, you will lose more money by not spraying than you will spend to spray. Once the aphids break the economic threshold it is time to kill them with pesticide or incur significant financial loss.

There is an array of insecticides, both organic and conventional, to turn to, but aphids' main strength is their rapid reproduction rate, which allows them to quickly develop a tolerance to the insecticides used to kill them. If you repeatedly use the same insecticide, it will soon become ineffective against the aphids. If you are going to use an insecticide, whether it is organic or conventional, it is critical to rotate use of different classes of insecticides so that no one class is used more than once during the growing season. Again, you must make sure that the insecticide is labeled for use on hops in your state and follow the directions precisely. Avoid the use of broad-spectrum insecticides, and stick to those that have less impact on beneficial insects. Although we have chosen to use only organic insecticides such as neem oil, it is important to remember that many organic insecticides are nonselective and can do significant damage to beneficial insect populations. Without the beneficial insects, your efforts at controlling aphids will be greatly compromised.

There are a few other steps that can be taken. Hop aphids prefer to eat the parts of the plant that are highest in nitrogen. These are generally the areas where growth is occurring. Avoid overfertilizing the hops because the resulting heavy onset of new growth attracts aphids. To nip an aphid outbreak in the bud you can remove potential winter hosts growing in the vicinity of your hop yard, such as plum or cherry trees. Or if you can identify the location of the aphids' winter host, you can spray the host plants to kill aphids in the spring or fall. You will still be using an insecticide, but you will be covering a much smaller area and therefore using less of it. If the aphids are overwintering in a commercial orchard, the orchardist is likely spraying for them already, as they cause damage to stone-fruit tree buds and early leaves.

Japanese Beetles

Japanese beetles can cause a lot of damage in the hop yard if they get out of control. These colorful, metallic-looking beetles appear exotic, and in fact they are. In their native Japan they have plenty of natural enemies and do not pose a threat to agricultural crops, but here in the United States it is a different story. Japanese beetles are members of the insect family Scarabaeidae, which contains some thirty thousand different species of beetles worldwide. In 1916 the Japanese beetle found its way to the United States among some iris bulbs shipped to a New Jersey nursery from Japan. Since the Japanese beetle voraciously feeds on over three hundred different types of host plants, it had no trouble finding something to eat and promptly infested the area surrounding the nursery. It was quickly recognized to be a major threat to agriculture, and in 1918 the government attempted to eradicate the beetle by spraying arsenic to kill the adults and saturating grassy areas where the larvae develop with cyanide. It didn't work.

By 1924 the Japanese beetle occupied 500 square miles (1,295 square kilometers). The government quarantined the area, inspecting all agricultural products moving in and out of the contaminated zone—searching nearly 1 million packages and inspecting 17,000 carloads, 11,000 truckloads, and 400 boatloads of goods in the first year. In the meantime they tried everything to kill the beetle, including traps, fumigation with liquid cyanide, electrostatic fields, hot water treatment, and DDT. But nothing worked. The Japanese beetle continued to chew its way across the country and today resides in most of the states east of the Mississippi River as well as Iowa and Missouri. Isolated infestations have been reported in other states, including California, Colorado, Oregon, and Washington.

Unlike most major insect pests of hops (which discreetly use their sucking mouth parts to pierce the surface of the plant and suck out the sap, leaving behind tiny spots on the leaf surface) Japanese beetles,

Although they are only just starting to appear in the Northwest, Japanese beetles have proved to be a significant problem in hop yards in the East.

One way to get a jump on things is to use Japanese beetle traps. These are the little yellow-and-green contraptions that seem to hang in every suburban yard. They are readily available at your garden supply store. The traps are baited with a synthetic version of the pheromone female Japanese beetles emit to attract males. The traps are useful for monitoring the arrival of Japanese beetles. When they start to appear in the trap, you know it is time to start worrying. But be careful with the traps; there is a common misconception that the purpose of the traps is to trap and kill all the Japanese beetles that are after your plants—so people put them right beside the plant they are trying to protect. This is a mistake. Rather than killing all the Japanese beetles in the vicinity of the plant the trap will simply continue to lure more and more beetles. Only a small percentage of these beetles will actually end up in the trap. Most will end up on the plant you are trying to protect. When using Japanese beetle traps to monitor the beetles, be sure to place them outside of the hop yard, using them only to monitor the beetles' arrival or to lure them away from the plants you are trying to protect.

That being said, scouting for Japanese beetles is pretty easy because they are large and do a lot of obvious damage to the leaves. Scouting twice a week is sufficient to monitor their arrival, but once the Japanese beetles are in the hop yard the population can grow quickly and significant damage can happen fast. After Japanese beetles first appear, step up the scouting to every other day or daily so you can be aware of when damage levels become critical. Be careful not to mistake false Japanese beetles (yes, there is an actual insect by this name) for the real thing. False Japanese beetles, also known as spring rose beetles, emerge earlier and feed on the flowers of plants that blossom in the spring such as peonies, apples, and cherries. False Japanese beetles are not as shiny. One surefire way to tell them apart is they do not have those striking little tufts of white hair on their abdomens. Before you see the Japanese beetle itself the first thing you will likely notice are the jagged holes in the leaves where they have been feeding. As the feeding progresses the leaves will appear skeletonized, with all the vegetation in between the leaf veins missing. Japanese beetles begin to feed at the top of the plant and work their way down. By the time they are on the scene, though, the hop plants will likely be too tall for you to easily see damage at the top.

Japanese beetles are pretty slow and clumsy. If you only have one or two hop plants it might be possible to pick them off by hand. Of course, this is laborious when you consider the job must be done from top to bottom, and if your hops are full grown, the top can be up to 20 feet (6.1 meters) in the air. If you don't want to climb to reach the top of the bine, a good number of beetles can be brought down by shaking the string the bine is climbing. Beetles plucked by hand can be easily and safely killed by dropping them in a bucket of soapy water. You can also kill the beetles by spraying the bine with soapy water or using an insecticidal soap, also available in organic form or easily made at home. To make a base for an insecticidal soap, mix 1 cup (240 milliliters) of vegetable oil with 1 tablespoon (15 milliliters) of Castile liquid soap. Blend the base with warm water at a ratio of 2 teaspoons (10 milliliters) of base to 1 cup of water.

Avoiding synthetics, we have had pretty good results in our small hop yard controlling Japanese beetles by spraying neem oil on the plants. Neem is an evergreen tree that grows in India. Neem oil, which contains a toxin called azadirachtin, is a naturally occurring pesticide that has been used for centuries. It is extracted from the tree's crushed seed. Neem oil functions as both a repellent and an insecticide. Its presence on the plant drives Japanese beetles away, and if they do end up ingesting neem oil, it will kill them over time. Neem oil is approved for use on organic farms, and it breaks down quickly in the environment. There are several commercial brands of insecticide that contain neem oil as well as synthetically produced azadirachtin.

Pyrethrin is another organic insecticide that you can spray on hops to kill Japanese beetles that is approved in New York and New England. You do have to be careful when spraying pyrethrin because it kills beneficial insects as well. Killing off beneficials can lead to more problems by causing what are called secondary outbreaks. A secondary outbreak occurs when the population of another pest, one that was being kept in check by the beneficials, takes off when the beneficials die after you sprayed for the original pest you were trying to control.

There are several other more toxic options for the control of Japanese beetles. Before using one you have to make sure that it is labeled for use on hops in your state and also consider the possible consequences for the environment and the safety of humans and wildlife.

Aside from spraying insecticides there are other steps that can be taken to control Japanese beetles. Kaolin, a soft, white powder made of the clay mineral kaolinite, is one of these options. Kaolin, which is mined around the world, is used to make ceramics as well as rubber and paint. When it is mixed with water it forms a slurry that can be sprayed on plants. It does not function as an insecticide but instead works more like a repellent. The clay dust gets into the Japanese beetle's joints and acts as an irritant, forcing it to go elsewhere to feed.

Another less toxic option is sprinkling diatomaceous earth on the ground at the base of the hop plants. Diatomaceous earth is a very soft type of rock that can be crumbled into a powder. It is not poisonous but kills insects by getting on the surface of their bodies as they crawl in the dirt. The powder then absorbs liquids from the insects, causing them to dehydrate.

Although Japanese beetle larvae cause a lot of trouble for growers of turf and sod, they do not cause damage to the hop plant. However, their presence is a clear indicator of how big a problem you may have the following season. Of course you can't see the larvae because they are underground. You can scout for them by looking for patches of dried-out grass. You can roll back these patches of dead grass cover and look for the grubs, which by September will be about 1 inch (2.5 centimeters) long. They will likely be at the edges of the dead area seeking out roots of the green grass on which they feed.

The larvae can be killed by releasing beneficial bacteria or nematodes into the ground. Nematodes are microscopic worms that carry bacteria that cause disease in larvae and can be purchased from any garden supply source. One nematode called *Heterorhabditis bacteriophora* that is effective against Japanese beetle grubs is available commercially and goes by the trade name B-Green. Nematodes are mixed into water and applied directly to the soil as a spray at times when the soil is already moist and the air and soil temperature are cool.

Paenibacillus popilliae is a bacterium naturally occurring in the soil that causes a disease of Japanese beetle larvae called milky disease. It only affects Japanese beetle larvae and is not toxic to any other life forms. This bacterium is available commercially as a powder under the trade name Milky Spore. It was the first microbial insecticide to be registered in the United States, back in 1948. It is applied to the soil as a dust and only harms Japanese beetle larvae. This may sound like an ideal fix, but there are some challenges. Milky Spore application can only be effective if soil temperatures are between 66 and 70 degrees Fahrenheit (18.9 and 21.1 degrees Celsius). For many parts of the Northeast cooler temperatures will slow the spread of Milky Spore through the soil, and it may take several years before it starts to make a difference. Low levels of Milky Spore in the soil are not effective because the larvae must actually eat the spore to become infected. However, once established Milky Spore can stay in the soil, helping to control the Japanese beetle population for as long as fifteen years.

Bacillus thuringiensis, which is commonly referred to as Bt, is another bacterium that can be used to control Japanese beetle grubs as well as many other types of larvae. Upon entering the larva the bacterium produces a protein that paralyzes the larva's digestive system, causing it to starve to death. Bt is found in soils throughout the world. It was developed for use as a microbial insecticide in the 1920s. Because it is naturally occurring and toxic only to specific insects, it is frequently used by organic farmers.

Taking a step up from the microscopic world beneath the surface of the soil, there are numerous excellent predators of Japanese beetles flying in the sky around your hop yard—birds. Not only do birds eat insects, many birds—particularly crows, grackles, robins, and starlings—troll the ground in search of grubs. In addition mammals such as moles, skunks, and opossums will also dig up grubs and eat them. Encouraging birds and wildlife that eat insect pests is another reason to keep your hop yard free of toxic chemicals.

Here is a potato leafhopper adult and a nymph on a hop leaf. The potato leafhopper punctures a vein of the hop leaf with its sucking mouthparts to remove sap. While feeding, the leafhopper injects a toxin into the leaf vein that causes a blockage, preventing the flow of nutrients. Photograph courtesy of UVM Extension Northwest Crops and Soils Team

Potato Leafhopper

The potato leafhopper, *Empoasca fabae,* is a major agricultural pest in the Eastern and the Midwestern United States. Leafhoppers are a big family of insects called Cicadellidae, which feed on plants and are found throughout the world. Leafhoppers are tiny and quick-moving, known for their powerful hind legs and wings, which allow them to move quickly. One of the biggest crops they impact is alfalfa, but they are also found on clover, potatoes, beans, raspberries, and strawberries. The leafhopper is turning out to be a pretty major pest of hops in the East. Like the hop aphid it harms the hop plant by sucking out its juices—but unlike the hop aphid it travels great distances to do so.

LIFE CYCLE

Potato leafhoppers cannot survive cold temperatures, so they overwinter in the south, mostly around the Gulf of Mexico. In the spring these audacious little bugs boldly hitch rides north on storm fronts forming in the Gulf. The low pressure system associated with a storm prompts adult potato leafhoppers (usually females carrying fertilized eggs) to fly upward, where they catch an updraft and are drawn high into the clouds. They travel north with the storm, then ride downdrafts to the ground, where they look for something to eat. The potato leafhopper can make the complete trip north in five days or fewer.

Adult potato leafhoppers are tiny, bright green, wedge-shaped insects. Their bodies are broadest at the head and taper down to their closed wingtips. With a magnifying glass six white spots are visible on their backs just below their heads. They earn their leafhopper name by walking quickly backward, sideways, and forward, leaping and flying. The nymphs look much the same as the adults but are smaller and without wings, yet just as speedy.

Potato leafhoppers start turning up in the hop yard in May. Their numbers grow slowly during the spring. Not only do they begin to reproduce in the hop yard, new potato leafhoppers continue to arrive in greater numbers through June and July. They do

At first, feeding by the potato leafhopper causes the veins of the hop leaf to begin to pale, then the leaves start to curl. The tip of the leaf will yellow and eventually crinkle and turn brown in a characteristic V-shaped pattern called "hopper burn."
Photograph courtesy of UVM Extension Northwest Crops and Soils Team

the most damage from mid-June through mid-August. Although they can survive until frost, toward the end of the summer the adults stop reproducing, and their numbers begin to drop.

Like the hop aphids, potato leafhoppers use their sucking mouthparts to suck sap from the plant by penetrating the veins in the leaves. To make matters worse there is a toxin in the potato leafhopper's saliva that forms a blockage in the leaf vein, depriving the hop of nutrients. The insect itself is so small that you may not know it is there until you begin to see the damage it is causing. At first the veins begin to pale, then the leaves start to curl. The tip of the leaf will yellow and eventually crinkle and turn brown in a characteristic V-shaped pattern, called "hopper burn." Hopper burn not only prevents nutrients from flowing through the plant but compromises the plant's ability to photosynthesize sunlight, which is how it gets its energy. If the plant is severely impacted, its internodes (the sections of stem in between the leaves) will be shortened, the plant itself will become stunted, and it will not produce many hop cones.

Potato leafhoppers must be monitored closely because shortly after arriving they begin to lay eggs, and their numbers can build quickly and even explode. The fertilized females kick things off by laying two to three clear, gelatinous eggs on the stems of leaves. The females will continue to lay two or three eggs a day for at least a month. The eggs, which are invisible to the naked eye, hatch in seven to ten days. Nymphs feed in the same manner as the adults, primarily on the undersides of the leaves. They go through a series of five molts over a period of two weeks until they reach adulthood. The period of time between when the egg is laid and when the nymph reaches adulthood and can begin to reproduce is twenty-six days.

PREVENTION AND CONTROL

Although potato leafhoppers are tolerable in the hop yard at low numbers, if the population explodes they can cause a lot of damage. Unfortunately, at this time there are no hop varieties bred to tolerate potato leafhoppers; however, the University of Vermont is currently conducting research into which varieties can stand up to them the best. In terms of cultivation practices the only way to control potato leafhoppers without the use of insecticides is to encourage the presence of predatory insects in the hop yard.

It is important to start scouting for potato leafhoppers in May when they begin to arrive. Because they come in on storms it makes sense to look for them after storms pass through the hop yard. When their numbers are low they can be hard to spot. One way to find out if they are arriving is to hang bug traps (essentially the glue-covered cards you can buy from garden supply centers) to trap bugs in your hop yard so that you can monitor the arrival of the potato leafhoppers. The wind will blow the potato leafhopper, along with any other insects in the vicinity of the card, into the card, where it will be trapped by the glue on the card.

Because potato leafhoppers are not a problem in the Northwest, there is no established IPM economic damage threshold for these insects on hops. But UVM recommends hop growers adopt the threshold set for alfalfa, raspberries, and potatoes, which is two to three potato leafhoppers per leaf. Once a week scout twenty-five to thirty plants scattered throughout the hop yard. Look at the undersides of two to three leaves per plant, and keep track of how many potato leafhoppers (adults or nymphs) you see on each leaf. Then calculate the average number of potato leafhoppers per leaf by adding up the total number of insects you sighted and dividing that by the total number of plants you inspected. If this number turns out to be larger than two, you may want to consider spraying an insecticide, preferably organic.

Alfalfa is one of the potato leafhopper's favorite food sources. If your hop yard is near an alfalfa field, which is likely if you are in an agricultural area, you will want to be even more vigilant. Pay attention to what the farmers in your neighborhood are doing. Once an alfalfa yard is cut, the potato leafhoppers that were living there will be on the move looking for a new food source. They may choose to relocate to your hop yard. When alfalfa in your area is being cut, you should scout for potato leafhoppers at least twice a week.

Because these insects are very small and fly away quickly when disturbed, it can sometimes be hard to get an accurate reading of how many are actually on the leaves. Alfalfa growers scouting their fields sweep an insect net at various locations throughout the hop yard. It is still hard to count the leafhoppers you capture, because leafhoppers are so nimble they will jump right out of the net as soon as you open it. For the most accurate reading you should put the insects you've collected into a kill jar, then sort them out after they are dead. Researchers such as Lily Calderwood at UVM also use a vacuum trap to suck insects off plants and into a kill jar.

There are several different types of insecticides that can be used to kill potato leafhoppers. The *2014 Cornell Integrated Hops Production Guide* recommends the use of pyrethrin, a botanical pesticide derived from chrysanthemum flowers that impacts nerve function, paralyzing and killing insects. It is considered safer than organophosphate pesticides, which are extremely toxic to humans as well as birds and other mammals. Pyrethrin is approved for organic production. Synthetic versions of pyrethrins are called pyrethroids.

A word of caution: do not overreact to a potato leafhopper invasion and spray an insecticide unless you absolutely have to. Insecticides used against potato leafhoppers, such as pyrethroids, will kill off your predatory mite population. A strong predatory mite population is an essential partner in controlling the two-spotted spider mite, which usually comes on strong in August. The two-spotted spider mite is a more serious pest in the hop yard because it causes damage to the hop cones close to harvest.

Two-Spotted Spider Mite

The two-spotted spider mite, *Tetranychus urticae,* also known as the red spider mite, is such a regular in the hop yard that it is often referred to by hop growers by the acronym TSSM. Mites are very small arthropods. Some are microscopic. Barely visible, a female two-spotted spider mite is only about 1⁄50 of an inch long, with a male being slightly smaller. The TSSM spins a web, hence its name. But the webs don't look like spiderwebs. Instead they are a sort of matted layer of whitish webbing just above the surface of the leaves, sort of like a tarp the mites create to protect them while they feed below.

TSSMs eat hundreds of different kinds of plants and are considered a major agricultural pest. TSSMs are oval-shaped and yellow to yellow-green in color, with dark spots on both sides of their abdomens, where the food they have consumed is visible inside their bodies. In late summer and early fall they darken to an orangish color. The adult mites have four pairs of legs, while the newly hatched larvae have only three pairs. Eggs are tiny, clear-to-pearly-white spheres only 1⁄200 of an inch in diameter. Adult TSMMs are normally found on the underside of leaves, where they feed on leaves and cones by puncturing the plant's surface and sucking out the plant juice, causing the vegetation to weaken and bronze. They reproduce rapidly, and when their numbers become high they cause defoliation. Their feeding on cones causes the cones to turn red in color and become dry and brittle, rendering them useless.

LIFE CYCLE

Understanding the TSSM's life cycle is essential to controlling it. Taking shelter in protected areas such as hop crowns and cracks in hop poles, dormant females overwinter right in the hop yard. They emerge in the early spring, beginning the season by feeding on hop shoots, getting things off to a quick start. The females begin to lay eggs as early as two days after they come out of hiding, and the eggs hatch just a few days later. Female mites hatch from fertilized eggs and males from unfertilized eggs. The larvae become fully mature in three weeks or less depending on the temperature; upon maturity, they begin to breed. More eggs are laid and so on. Females can lay up to 16 eggs a day and as many as 240 eggs during their lifetime. Obviously things can get out of hand quite quickly.

PREVENTION AND CONTROL

TSSM infestations become worse in hot, dry weather. Dusty conditions also are favorable for the mites because it makes it easier for them to get traction on the leaves. Because of this, TSSMs are a major issue in hop yards in the Yakima Valley. Although hop vegetation can withstand a fairly high level of TSSM infestation, a heavy presence of mites will severely impact the quality of the cones, so it is important to try to keep the population under

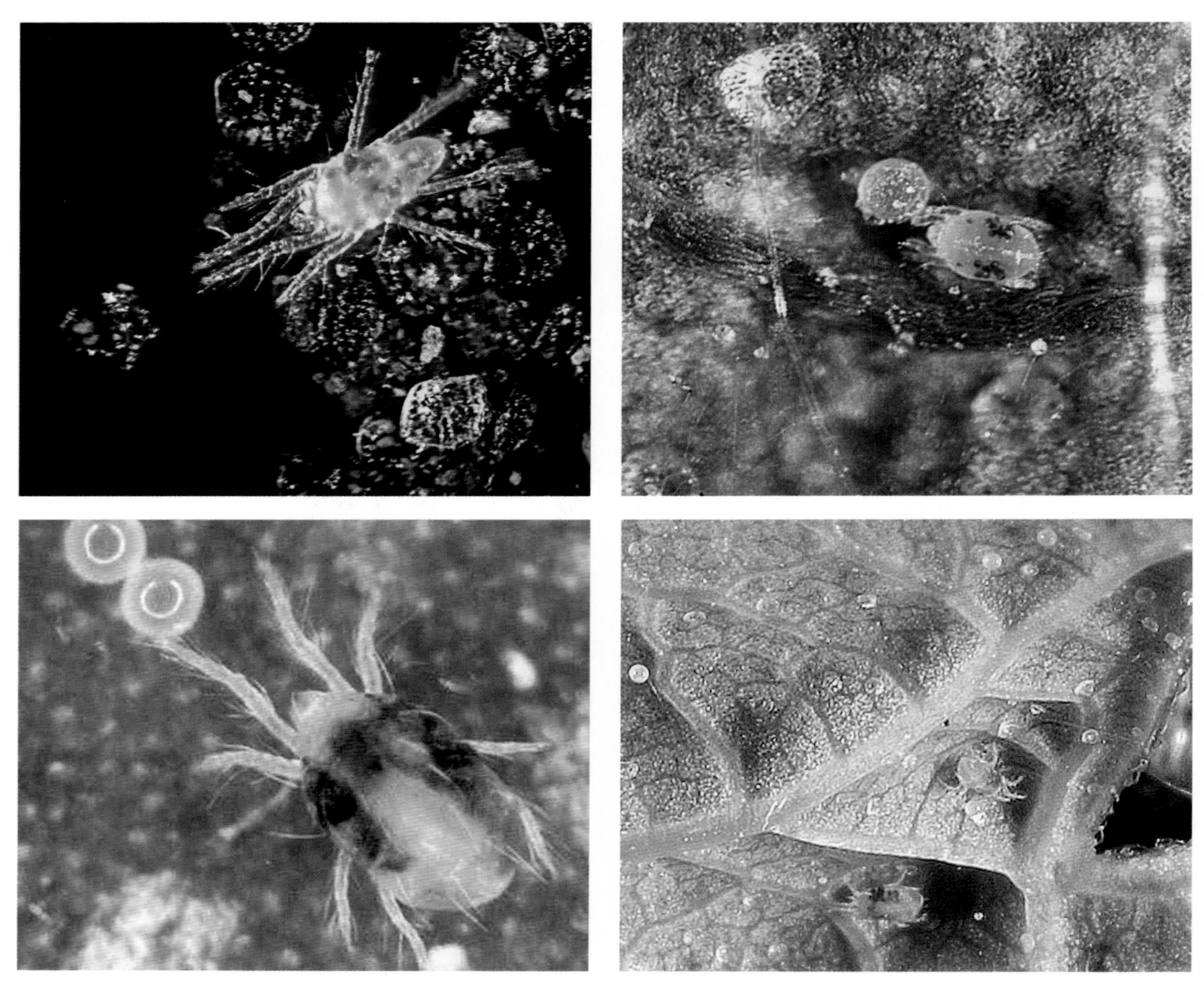

The two-spotted spider mite is a major pest in hop yards. Since it thrives under hot, dry conditions, it usually becomes a problem in late summer. Photographs by D. H. Gent, *Compendium of Hop Diseases and Pests*, American Phytopathological Society

control before it gets to that point. Using magnification, start to check the undersides of hop leaves for TSSMs in May. Examine the leaves at a height of between 3 and 6 feet (0.9 to 1.8 meters). Look for mites as well as eggs, webbing, and the telltale stippling and yellowing of leaves that indicates their feeding. As the hop bines grow, the mites move onto the growing tips of the bines. In mid-June use a stepladder or pruning pole to collect leaves from the top of the bine and inspect them for signs of mites. Growers generally begin to spray when they see between one and two mites per leaf in June and five to ten mites per leaf in mid-July.

Before spraying an insecticide, it is particularly important with TSSM to stop and think about the beneficial insects. Scout for them while you are monitoring the TSSM population. Predators of TSSM include predatory mites, big-eyed bugs, pirate bugs, ladybird beetles, spiders, and lacewings. To control diseases such as downy mildew, hop growers are advised to keep the areas around the hop plants free of weeds, but vegetative cover

Two-spotted spider mites pierce the plant to suck out juices, leaving a stippling pattern, ultimately causing the leaf to turn yellowish in color. Photographs by D. H. Gent, *Compendium of Hop Diseases and Pests*, American Phytopathological Society

actually supports populations of beneficial insects. Including plant species that attract beneficial insects in your row ground cover planting can also help build the number of beneficial insects (see the cover cropping information in Chapter 4). Providing adequate water is also important. If conditions are too dry, not only does this weaken the population of beneficial insects, drought stress raises the temperature of the hop leaves, a situation favorable to TSSMs that enables them to reproduce even faster.

If the TSSM population in your hop yard reaches a level at which you feel the application of an insecticide is warranted, choose your weapon carefully. Make sure that it is "selective," targeting mites specifically. You want to avoid killing all the beneficial insects that prey on them. Not only are these beneficial insects good for the hop yard in general, they are essential to the ongoing control of TSSM. If you spray a "nonselective" pesticide and kill both the TSSM and its predators, because of its rapid reproduction rate the TSSM will recover from the

Two-spotted spider mites have fed on these cones, ruining them. Photographs by D. H. Gent, *Compendium of Hop Diseases and Pests*, American Phytopathological Society

insecticide application much more quickly than the beneficial insect population will. Without the presence of that natural control, the TSSM population will take off again and quickly become worse than it was before you sprayed. That is what is called "flaring." Flaring is one of the biggest challenges in the control of TSSM. It is important to remember that just because an insecticide is organic does not mean that it is "selective."

Also, remember that the first line of defense against all insect invaders is your plants themselves. Make sure you have healthy, well-nourished hops, cover crops, and soil, and develop a hop yard ecosystem that works with, not against, nature.

CHAPTER 10

Weed Control

It is one of the great ironies of agriculture that as you struggle to successfully grow your crop of choice all kinds of other plants in your field seem to have no problem at all thriving, to the point at which they threaten the very existence of the plant you are trying to grow. In general all these other plants are known as weeds. By its very definition a weed is simply a plant growing somewhere it is not wanted. In general weeds are plants that grow wild, spread aggressively, and interfere with the growth of a cultivated plant.

The effects of climate change aren't going to make the battle against weeds any easier. Research is showing that the increasing carbon dioxide level in the atmosphere and rising temperatures we are experiencing due to climate change are working against us when it comes to weed control. As the temperature increases and frost-free dates are pushed back, the growing season for weeds is extended. Temperature changes also increase the risk of invasive weeds moving up from the south. At the same time, rising levels of carbon dioxide in the atmosphere appear to make some weeds grow faster and enhance their reproduction.

People have been at war with weeds ever since they first began to farm, over ten thousand years ago, and during this time have developed many strategies for controlling them—from simply pulling them up by hand to using heavy machinery to blast them with highly toxic synthetic chemicals. When weeds are out of control in a hop yard, they compete with the hops for water as well as nutrients from the soil. When the hops are still small, weeds also compete with them for sunlight. Weeds are a particular problem in hop yards because they provide a cover for harmful insects and also create a shaded, humid environment around the base of the hop plant, which is conducive to the development and spread of fungal diseases such as downy mildew, as well as other disease pathogens. Controlling weeds in the hop yard is difficult because the perennial hop plants are permanent fixtures. Once the hops are in the ground the soil in the hop yard cannot be turned entirely over to kill weeds or sow a weed-suppressing cover crop. This is one reason it is important to take major steps to control weeds before the hop rhizomes are planted. That being said, no matter what extremes you go to before the hop yard is planted you are still going to be dealing with weeds, and there are a variety of methods of control.

Prevention

As always, prevention is the first step, and some of the advance planning you will do in your hop yard will be geared around future weed control—as described in previous chapters. Planting cover crops to choke out existing weeds in your hop yard site before planting rhizomes is a great way to get started, as discussed in Chapter 4. But in their efforts to spread their species far and wide, weeds have ways of finding their way into your field, and in many cases

Too many weeds in the hop yard not only compete with the hops for nutrients and water, they provide an environment conducive to the development of disease pathogens and insects.

there is little you can do about it. Weed seeds fly in on the wind as well as in the stomachs of birds and other types of wildlife. If your hop yard is healthy habitat you will have plenty of critters moving in or passing through, and they will bring weed seeds with them. There's not much you can do about any of that.

You can, though, take measures to prevent seeds traveling into the hop yard on the clothes and shoes of human visitors or on farm equipment. The same ad hoc biosecurity measures recommended for preventing the spread of disease (see Chapter 8) apply: bring in only clean plants, wash equipment, and change clothes as necessary.

You should also be aware, though, that weed seeds can be brought in with irrigation if you are using surface water as a source. And they can also flow into the hop yard with runoff from surrounding fields during heavy rain. So it's not enough to simply prevent and control weeds inside the hop yard; you must also prevent and control them around the borders of your yard and around your water source.

CONTROLLING WEEDS WITHOUT HERBICIDES

In general the best weed management plan for a hop yard involves heavy mulching around the bines in the spring to block weeds from growing up in and around the hop, keeping ground cover in the aisles of the yard mowed to avoid plants going to seed, whacking back any weeds that break through the mulch barrier, and manually removing any weeds that infiltrate the crown.

But the better you understand weeds, the better you'll be able to control them. Basically there are two categories of weed—broadleaf weeds and grassy weeds. Broadleaf weeds, such as ragweeds, have flat leaves that grow horizontally. Grassy weeds, such as crabgrass, have bladelike leaves that grow vertically. It is important to differentiate between the two because not only do they grow differently, the strategies you will use to control them will differ as well.

It is also key to understand whether a weed is perennial, biennial, or annual. A perennial is a plant that lives for multiple years, a biennial plant has a two-year life cycle, and an annual plant completes its life cycle in one year. There are two types of annuals—summer annuals and winter annuals. A plant that is a summer annual emerges from the soil in the spring, grows and produces seed in the summer, and is killed by frost in the fall. Summer annuals include lamb's-quarter, purslane, and pigweed. Winter annuals germinate in late summer or fall, stay alive through the winter, and flower and produce seed in the spring. Winter annuals include shepherd's purse and chickweed.

Reducing weeds in the hop yard can be done by planting cover crops to choke out weeds prior to installing the hop yard and using mulch—such as plastic, landscape cloth, or composted wood chips or bark—around the base of the plants.

When weeds do arrive they are virtually impossible to eliminate, but they must be kept under control. What type of control works best depends on how the weed reproduces itself. Annuals and biennials (and also, in some cases, perennials) reproduce by flowering and going to seed. To control the spread of seed from existing weeds, the trick is to stop this from happening. Most weeds produce a huge amount of seed, and much of this seed can remain viable for a very long time. Every time a single weed in your hop yard goes to seed your problem gets significantly worse. For example, the best and cheapest way to control broadleaf summer or winter annuals—which have shallow root systems and reproduce themselves through seed production—would be to simply pull them out of the ground before they go to seed. That approach is less feasible when dealing with a patch of a perennial grassy weed that enlarges its territory by expanding its deep underground root system. In such cases, you'll need to employ other methods.

It is easy enough to mow the aisles in the hop yard so the plants growing there never have the opportunity to go to seed. In the rows of hops themselves, you can carefully wield a weed whacker between the bines. But for weeds growing close to the base of the hop, a hoe is a better bet so that you don't risk cutting the bine. When weeds infiltrate the hop crown, and they will, they must be pulled by hand. The bigger your hop yard, the more time-consuming this work becomes; but it is worth it in the long run.

Ragweed is an annual, broadleaf weed with shallow roots that grows to a height of about 4 feet (1.2 meters). Ragweed seeds in the soil germinate from May through early June. The plant flowers in August through early September, producing as many as a billion grains of pollen per plant. This pollen spreads far and wide on the wind, pollinating other ragweed plants and setting off hay fever attacks in humans. (Interestingly, seasonal hay fever is an allergic reaction not triggered by hay, but instead by mold spores and pollen from grasses, trees, and plants such as ragweed.) But for farmers the next step is even worse than the pollen release. Once pollinated, the plant disperses a horrifyingly large number of seeds. One ragweed plant can produce as many as 69,000 seeds. To make matters worse, ragweed seeds can remain viable in the soil for more than thirty-nine years.

Ragweed is trouble in the hop yard because it is an extremely aggressive plant that will compete with the hops for nutrients. A severe ragweed infestation can cause nutrient deficiencies in hops. Ragweed can be suppressed in the hop rows with mulch. Because the plants have a shallow root system, it is pretty easy to pull any ragweed plants that manage to surface through the mulch, and young ragweed plants will not survive chopping with a hoe. Mowing the aisles between rows to cut down ragweed before it goes to seed is important, too; but remember that ragweed plants cut in midsummer will have time to grow new stems and still be able to flower unless cut again.

If the presence of ragweed in the hop yard is overwhelming, and not responding to nonherbicide measures, ragweed can be controlled through the application of most types of preemergent herbicide targeting broadleaf weeds, following up with a postemergent herbicide application to kill any survivors. Unfortunately, there are not a lot of effective organic options for preemergent herbicides. Avenger is a citrus-based organic postemergent herbicide that is quite effective. Unfortunately, it is not selective so don't get any on your hops. Ragweed woes have prompted many to consider Roundup, which has historically been used to kill it, but incidents of ragweed developing resistance to Roundup have been documented by the International Survey of Herbicide Resistant Weeds (on farms growing crops such as cotton, corn, and soybeans) all around the country over the past ten years.

Because ragweed crops up where people do, and many people are desperately allergic to it, there has been a little more attention to developing environmentally friendly means of control. This need has been exacerbated by climate change, which is working to ragweed's advantage. The plant benefits from increased carbon in the atmosphere, which enables it to grow faster and produce even more pollen. In addition, climate change has pushed back the hard frost dates in many regions, prolonging the plant's pollen production and along with it the ragweed allergy season. Hopefully, outcomes of this research will also benefit farmers who prefer not to use synthetic herbicides.

FIELD BINDWEED

Field bindweed, *Convolvulus arvensis*, is an interesting weed to combat in the hop yard because the qualities that make it "bad" as a weed are the same qualities that make hops strong and resilient. Field bindweed is a broadleaf, hardy perennial vine that resembles a morning glory. Native to Eurasia, it invaded the United States via Virginia back in 1739, possibly as an ornamental plant. It worked its way across the entire eastern United States by the early 1800s. Today it is considered one of the most problematic weeds facing agriculture located in temperate regions worldwide. Belowground, field bindweed's main root reaches a depth of 20 feet (6.1 meters). This vertical root then sends out lateral roots; once they reach a distance of 15 to 30 feet (4.6 to 9.1 meters) away from the parent plant they turn downward and become a secondary vertical root, which sends up its own shoots. Like the hop, field bindweed also has underground rhizomes that produce buds and form new plants. Just to make

Field bindweed is a perennial vine that, much like the hop, spreads through an underground system of roots and rhizomes.

sure it has all its bases covered, each field bindweed plant will also produce and spread over 550 seeds, each of which can remain viable in the soil for up to sixty years and can germinate in conditions ranging from 41 to 104 degrees Fahrenheit (5 to 40 degrees Celsius). Once aboveground, field bindweed grows along the surface until it finds something to climb, usually another plant, which it entwines and eventually envelops. In other words, field bindweed is a tough one to beat.

Conventional means of mechanical weed control such as hoeing and tilling only make matters worse. Chopping up its belowground system of roots and rhizomes only helps it spread faster; however, continuously removing the surface vegetation through hoeing or mowing does have the effect of forcing the plant to use up its carbohydrates stored belowground. The plant will eventually starve to death if you keep this up.

One of the best ways to control field bindweed, though, is to block its light source. This technique works well in the hop yard since hop rows are often

Bindweed in the hop yard can be a particular problem, especially when the bindweed starts coming up in the middle of the hop crown and hop and bindweed shoots and roots become entangled.

heavily mulched as a matter of practice. When dealing with bindweed, nonorganic mulch such as black plastic, which can completely block the sun, is most effective. But you must be vigilant, as field bindweed searching for light has been known to actually puncture plastic row covers. Certainly the bindweed will emerge out from under the edge of the plastic and backtrack across it in an attempt to climb the hop. It will be need to be cut back with a weed whacker or other implement before it can work its way over to the hop plant.

Herbicides such as glyphosate work on field bindweed, but multiple treatments will be necessary and therefore extremely bad for the environment. Bindweed is a significant problem in our hop yard. Although we have not eliminated it, we have suppressed it by laying down black plastic mulch in the hop rows, keeping the hop aisles mowed, and weed whacking where the mulch layer meets the mowed vegetation in the aisles—where bindweed has a tendency to creep out. Of course it will periodically emerge within the crown, and at this point it has to be pulled by hand. So far we have not had to spray herbicide.

Pigweed grows fast and can reach a height of 7 feet (2.1 meters), competing with hops for sunlight and nutrients.

PIGWEED

Like ragweed, pigweed is a fast-growing, native broadleaf annual. Part of the grain-producing Amaranthaceae family, pigweed is actually the common name for a group of weeds that includes redroot pigweed, smooth pigweed, powell amaranth, waterhemp, prostrate pigweed, and tumble pigweed. It is an aggressive plant that thrives in hot weather. Seedlings sprout in great numbers in the late spring and early summer. Pigweed grows up to 3 inches (7.6 centimeters) a day and can reach a height of 7 feet (2.1 meters), easily choking out crops. The plant has the advantage of being edible—the young leaves can be served as a green fresh or steamed—which provides a perk when weeding.

A plant will begin to flower and produce pollen as soon as six weeks after emergence. Each plant produces thousands of flowers, and a single plant can produce between 100,000 and 600,000 seeds. Since it is a nitrogen-loving plant it can be a particular problem for hop growers, who must fertilize their hop yards heavily to reach production goals.

The best strategy for controlling pigweed is to strike early. Since its seeds are very small they do not pack a lot of nourishment for emerging plants. Because of this, pigweed seedlings need to get aboveground and start photosynthesizing to feed themselves. The large numbers of tiny seedlings that sprout up are so delicate and lightly rooted in the ground you can practically scrape them away with your fingers and they won't regenerate. However, if you don't get them when they are little, they grow fast and are harder to kill once established. It is best to control them by pulling or mowing before they flower, as they are quick to adapt to herbicides.

QUACKGRASS

Quackgrass is a perennial grass, and perennial grasses are particularly difficult to control. They are frequently a problem in hop yards, as fields not previously intensely cultivated, such as pastures where such grasses are well established, are often the first choice when it comes to locating hop yards. Quackgrass, native to Europe, has a lot of names, including couch grass, dog grass, quickgrass, quitch, scotch, twitch, and witchgrass. Such an abundance of names usually indicates a widespread problem.

It is an interesting challenge to have in the hop yard because, like the hop, it reproduces itself through rhizomes. An extensive underground system of roots and rhizomes spreads underground, sending up new shoots that can survive on their own if they are broken off from the parent system by mechanical cultivation. Aboveground the grass can grow to a height of between 1 and 4 feet (0.3 to 1.2 meters). The green grass generates nourishment for the underground rhizomes. As the end of the growing season approaches, the rhizomes store up energy so that they are ready to go in the spring.

Controlling quackgrass is a matter of combining tilling and mowing. Digging up the roots and repeating the process again before the rhizomes have time to resprout is one strategy. You can also turn over the soil and expose the roots and rhizomes to the sun in the height of summer, which will kill them. Back up

Quackgrass is a persistent grassy weed that reproduces itself through an underground root and rhizome system.
Photograph courtesy of Sid Bosworth, University of Vermont

the disruption of the root system with repeated mowing of the grass aboveground, keeping it as short as possible. This treatment reduces quackgrass's ability to generate and store enough energy to get it through the winter. After using these techniques during the growing season, plant a cover crop in the fall that will compete with the quackgrass for space. After a couple of years of this cycle of control, the quackgrass should be significantly compromised. This of course must be accomplished prior to the installation of the hop yard and the planting of the hops. If quackgrass continues to grow in the hop yard despite your efforts, you can use the above-described technique of root disruption followed by repeated mowing in the row covers while heavily mulching the planted rows to deprive the grass of sunlight.

PART IV
HARVESTING AND PROCESSING

CHAPTER 11

Harvest

Hop-farming wisdom says that you can have friends help you pick hops by hand for the first two years but when the third harvest comes around you won't have any friends left. That's why in our third year we purchased a small mechanical hop-picking machine that had been recommended to us. The machine was delivered the week before we left for Washington and Oregon on a learning tour of large commercial hop farms. This trip timing was cutting it close with our own harvest, but we thought it was important to see the hop yards out west when the bines were at their fullest and witness the beginning of their harvest. We had a mile-long list of questions for the growers out there. What we found out was that for fledging hop farmers from New York like us, going to a hop grower in the Yakima Valley for advice

As the end of summer approaches, the hops are ready to harvest.

My father, Peter Ten Eyck, pitching in to handpick hops at harvest time.

Danny Hallman is the farm manager at Golden Gate Emerald Hop Ranches in Sunnyside, Washington, located in the Yakima Valley.

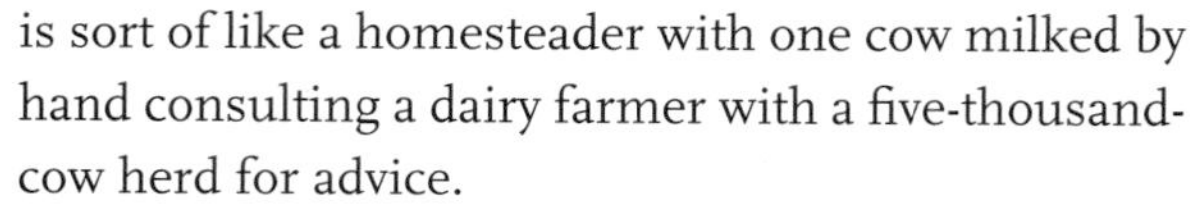

is sort of like a homesteader with one cow milked by hand consulting a dairy farmer with a five-thousand-cow herd for advice.

"One acre?" said an incredulous Danny Hallman, farm manager at Golden Gate Emerald Hop Ranches in Sunnyside, Washington. "You say you have 1 acre of hops?" We were standing in his air-conditioned office in Sunnyside, Washington, while outside industrial-sized Yakima combines and picking machines roared away on the first day of harvesting Emerald's 400 acres of hops. Hallman is a big bear of a man with a trimmed beard and bright eyes. Despite the huge gulf between our two farms in both geography and scale, he truly wanted to help us. So he asked how we went about harvesting our hops. We explained that we had picked hops by hand for the last two harvests. Under the visor of his baseball cap his eyes widened. Feeling somewhat idiotic, we went on to say that we knew this was unsustainable and had purchased a small mechanical hop-picking machine but had discovered just prior to leaving for this trip to the Northwest that it didn't pick hops cleanly off the bine but instead destroyed them. Hallman asked intently about the machine. Dieter explained to him that the machine had rubber fingers that raked the cones from the bine. "Rubber fingers!" Hallman said, aghast. "It has rubber fingers? That will never work. It is too much friction. The machine is going to have to be retrofitted with metal fingers. It's too bad the designers didn't talk to us first."

Apparently, there are lessons to be learned when it comes to adapting equipment for smaller scales of

Steenland Manufacturing, based in Roxbury, New York, had to rebuild the Hop Harvester 1000 we purchased from them after mechanical problems encountered during harvest drove us back to handpicking.

harvest, but the minds feeding the Eastern hops renaissance are at work on a host of innovations.

Harvesting Basics

Whether you have hundreds of acres of hops in the Northwest or a single acre of hops in the East, in general hops are ready to pick in late August. As with apples, different hop varieties ripen at different points throughout the season. The harvest concludes in mid-September. The central activity in the hop harvest is removing the hop cones from the bines. It sounds simple, but there are many ways of going about it, most of which are determined by the size of the hop yard. Any hops not being used right away for brewing, which in most cases are the vast majority, must be immediately dried and cured. In actuality it is the farm's drying and curing capacity that ultimately controls the pace of the harvest. In other words, don't get ahead of yourself and pick more hops at any one time than you have room to dry.

Since each hop variety is desirable to brewers for its unique characteristics, it is critical to keep the varieties separate during harvesting, processing, and packaging. When harvesting, start with the earliest maturing variety and work your way through the season to the latest maturing variety, picking, drying and curing, and packaging each variety in its turn. Each variety of hops has a traditional time that it is ready to pick, and growers with many varieties use these projections as a planning tool. However,

Hop bines are attached to hooks to be elevated and fed through the hop-harvesting machine.

from a track that runs along the ceiling. The track feeds the bines in between a set of rapidly moving vertical panels lined with metal fingers that pull the cones from the bines. The stripped bines emerge from the panels and are fed into a chopper. The chopped vegetation is then transported by a conveyor to a chute, which dumps it outside the building. The cones drop down onto another conveyor belt on which they bounce along as fans blast away shredded vegetation debris.

The conveyor belt moves the cones from the picking machine building into an adjacent, much quieter building housing the drying kiln. The drying kiln is essentially a large warehouse with two stories. The hops lie in a layer about 2 feet (0.6 meter) deep on the floor of the top story, which is lined with burlap cloth. On the floor below, propane heaters blast heat that rises up into the second story, drying the hops. Once the hops are sufficiently dry, the burlap cloth is rolled up and the hops drop onto a conveyor belt that takes them to a third building, where they are dropped into an enormous pile on the floor and allowed to cool to a uniform temperature. This process is called conditioning or curing and takes about twenty-four hours. The conditioned hops are then pushed through a hole in the floor, from which they drop onto another conveyor belt that takes them into a fourth building, where they are fed into a press and compacted into 200-pound (91-kilogram) bales wrapped in plastic cloth.

This process is basically the same on most commercial hop farms in the Northwest. Some farms sell their hops through a distributor, and others market them directly. To meet the needs of the range of breweries, these bales of hops can be sold as 200-pound bales, cut into smaller sections, and vacuum

Pulled from the bine by the harvesting machine's metal fingers, the hop cones bounce along conveyor belts.

Hop cones are delivered to the hop kiln floor for drying.

An Eastern-style top cutter—our friend Peter wielding a pole pruner and being hoisted on the forks of a tractor while safely ensconced in a metal cage that formerly held a 275-gallon liquid storage tote. Photograph by Laura Ten Eyck

packed, pelletized, or reduced to resin or oil before being shipped to the brewery.

While commercial hop farms in the East are struggling to grow, none has even remotely reached the scale of the Northwest hop farms and in fact are not likely to. Although many start-up hop farms in the East have spent a season or two picking their hops by hand, this quickly becomes impractical as the plants mature and the acreage expands. While backyard hop growers can easily pick their hops by

Calculating the Percentage of Dry Matter

So you think it might be time to pick your hops but you are not sure? You don't want to pick them too early because the chemical compounds in the lupulin won't have reached their optimal levels. You don't want to pick them too late because the quality of the lupulin will start to degrade. To get it right you need to know what the desired percentage of dry matter in the hop cones should be at harvest for the variety in question. Then you need to know when the cones have reached that point. To figure out the percentage of dry matter in the hop cones for the variety you are considering harvesting, conduct the following eight-step procedure. Warning: it involves a little math.

1. On a dry morning go into the hop yard with a ladder and harvest some cones. Pick ten cones growing at a height of about 16 feet (4.9 meters) from ten different sidearms and put them into a bucket.
2. Mix up the cones within the bucket so they are thoroughly blended (yes, it is kind of like the soil test).
3. Select two microwave-safe containers as close in size and material as possible that will hold fifty cones each. Weigh the empty containers separately, and record the weight of each.
4. Fill one container with the fifty hops, and weigh it when full. This is the "green weight." Write it down, and set aside the container of green cones.
5. Fill the second container with cones. Put this container of cones in the microwave set on low. Remove the cones every thirty seconds, and weigh them until they are 100 percent dry. You will know they are 100 percent dry when the weight stops going down. That weight is the "dry" weight. Write it down, and set the dry cones aside.
6. You now have both a green weight (determined in Step 4) and a dry weight (determined in Step 5). Subtract the weight of the container from both the green weight and the dry weight.
7. Divide the dry weight (less the container weight) into the green weight (less the container weight). The number you get is the percentage of dry matter that the cones in the sample had when you picked them.
8. Depending on the variety, hops are generally picked at between 20 and 24 percent dry matter. If the number you got is within that range, it is acceptable. The range of 20 to 25 is a general range. Individual hop varieties, such as Cascade, have a more precise range that falls within the general range. If your dry matter is lower than that, your hops are not ready to pick yet. If you want to be extremely precise, you can research the specific dry matter content at which a particular hop variety should be harvested. For example, Cascade should be harvested when between 22 and 24 percent dry matter.

Hop cones are ready to be harvested when they reach between 20 and 24 percent dry matter.

hand and air-dry them, harvesting hops on the scale necessary for commercial production, even at a small scale, requires an array of mechanized equipment ranging from the machines that cut down the bines to picking machines that pluck the cones from the bine, drying kilns, hop balers, and pelletizers. Options for purchasing small-scale mechanical hop pickers and drying equipment in the East, initially nonexistent, are now limited but are starting to expand. In some cases farmers with the mechanical know-how are building their own equipment, and UVM (see the "Build Your Own Hop Harvester" sidebar on page 213) has built a mobile harvesting machine and dryer, making the instructions available online (see Resources).

When to Pick

Picking hops at the right time is the first and most important activity of the harvest season. Pick hops prematurely and the important alpha and beta acids in the lupulin will not be fully developed, compromising the hops' ability to add the desired flavor and aroma to beer and reducing their value. Pick them too late and not only will the alpha and beta acids have degraded but the cone itself will brown, and upon handling it will crumble into hundreds of loose, papery leaves scattering the lupulin like dust.

To determine when to pick, you must evaluate the hop cones' moisture content. Although in general hops feel somewhat dry and papery when you touch them, they will become more so as the season progresses. If you pinch it between your fingers, a fully mature hop cone will feel dry and somewhat springy. Inside the cone, the lupulin will appear deep yellow in color and will be so sticky that after touching it you simply won't be able to wash it off. If you touch a grain of lupulin to your tongue, it will be highly aromatic and bitter—but you won't taste the bitterness right away. Don't be misled into thinking you didn't get enough of a taste and try more. Wait a second, and soon your mouth will flood with hoppy bitterness. Sensory experiences aside, technically speaking hops are ready to harvest when they have reached a point at which their weight is composed of between 77 and 79 percent moisture. To put it another way, hops are harvested when the cones are on average 23 percent dry matter—the exact percentage point depending on the variety. For example, the Oregon Hop Commission recommends that Nugget be harvested when it reaches 23 percent dry matter and that Willamette be harvested at 20 percent dry matter.

In hop yards in Washington and Oregon, one of the first questions we put to hop growers as well as hop scientists was how they know when it is time to pick. The answer was always the same. Pluck a cone from the bine and tear it in half from top to bottom. If it easily splits evenly right down the strig (the footstalk, or stem, where the plucked cone once connected to the bine) it's picking time. Goschie Farm now relies heavily on modern technology to analyze moisture levels in hops before harvest, but

UVM Hop Harvest Moisture Calculator

Harvest readiness can also be measured in terms of moisture level. UVM has a handy online Hop Harvest Moisture Calculator on their Extension website. The calculator walks growers through the steps of weighing samples, asks them to enter their results, then automatically calculates the harvest moisture percentage, as well as the corresponding dry matter percentage. It also asks growers to enter a desired final moisture content and uses the data entered to calculate a target weight for their bagged hops (see Resources).

An old-fashioned way of telling if a hop cone is ready to harvest is to tear the cone down the middle. If it splits evenly along the strig it is dry enough to harvest.

Gayle Goschie said her parents always used the simple tear method to determine when the hops were ready to pick.

Although it is critical to pick hops at the optimum time, there are certain cases in which you are forced to pick hops early, usually in order to avoid a total loss. If certain diseases, such as downy mildew, or infestations of insects, such as aphids, get out of control and threaten the quality of the cones, it may be best to go ahead and pick the hops before they are destroyed completely. Approaching bad weather may be another motivation. It is impossible to harvest hops when they are wet, so if a long stretch of rain is coming and the hops are almost ready, it may be better to pick them a little early. This is particularly true if you are already battling diseases such as downy mildew, which really take off when the hops are wet. The quality of the hops may be compromised, but at least you have something. Also, in some situations, heavy wind or equipment failure may cause trellis collapse. It is extremely difficult to put a collapsed trellis loaded with hops back up, so it may be necessary to harvest the hops instead if they are nearly mature. Again, the harvest may be worth less money, but it is better than nothing at all.

In some cases hop cones mature extremely early. Goschie explains that this is due to weather conditions that extend the time between bud formation and flowering. When this happens the cones grow very large and mature earlier. Some growers choose to harvest early in this situation. However, Goschie Farms sticks to their traditional harvest schedule, believing that early harvesting is not good for the hop. According to Goschie, it does not allow the plant to photosynthesize as much as it needs to in order to store enough energy in the root system for the winter. For this same reason, most commercial growers do not harvest cones at all from first-year plants but instead let them remain on the bine to build up their roots.

Harvesting by Hand

Backyard growers and beginning hop farmers often pick hops by hand in much the same way it was done in the nineteenth century. In both hand and mechanical harvest it is traditional to cut the bine from the trellis before removing the cones. However, some home brewers who are very particular about their hops prefer to let the bine stand and use a ladder to harvest daily only the hops that appear to be at their prime. This of course is a painstaking process, and it is questionable if, without testing the hop either by tearing it open or conducting a moisture analysis, you can know for sure if one individual hop cone is more ready than another hop cone on the same bine. That being said, brewing beer is as much alchemy as science, and a home brewer should be encouraged to do whatever is necessary to make the best beer. There is one real advantage of picking hops off the bine this way: it leaves the bine in place until the end of the growing season. This allows the plant to store the maximum amount of energy belowground.

Megan handpicking hops from a hop bine still growing in the hop yard.

The more common approach is to cut the bine and bring it to a picking station, where the cones are removed by pickers or a machine. Since hand harvesting hops is not a cost-effective means of harvest, it is usually only undertaken on a relatively small scale by those who are not planning on a huge payoff at the end of the season but instead are experimenting with growing hops or planning to brew their own beer. Modern-day hop pickers are usually volunteers. In our case most of our volunteer hop pickers have been friends, family, and people from our community who are really interested in what we are doing and have offered to stop by to help. The contributions of all of these people to our early hop harvests have been huge.

However, once you are an established for-profit business, it is actually illegal under state and federal

Our friends hard at work picking hops off the bines harvested from our pilot hop yard.

law to have volunteer workers pick your crop (or help with anything else for that matter) except under very specific conditions. In fact, not long ago a vineyard in California was fined for having people pick their grapes in exchange for wine.

ADVANCE PLANNING

When hops are being hand harvested, advance planning is necessary—not only to line up the help you'll need but also to ensure that the cones are picked from the cut bines immediately (important) and that the drying process begins right after the cones are picked. Ideally, that planning starts *before* planting: if you do not plan on purchasing a picking machine before your first harvest, assess your capacity to raise volunteers, and don't plant more hops than you think you can recruit people to pick. It takes one person half an hour to pick the hop cones of a mature hop bine. An average hop bine produces about 8 pounds (3.6 kilograms) of fresh hop cones. Multiply the time it takes a person to harvest a hop bine by the number of hop bines in your hop yard and that will give you an idea of the number of person-hours you will need people to put in to handpick the hops. For example, it would take about 500 person-hours to handpick 1,000 hop plants. If all the hops were to be picked on the same day, that would mean we would need about 60 people to come and pick for 8 straight hours.

In addition, you should estimate the volume of picked hops you will be harvesting in a given day and make sure you have the capacity to dry them. Because you can't pick more hops than you have room to dry, limited drying space can slow down the picking process. It can be frustrating if you have mature hops on the bine, pickers on hand, and beer on tap but no more drying space. That's why it is important to think things through before you get going.

So during the summer start putting out the word, and line up your volunteers. When the hops are ready keep an eye on the weather, order a keg of beer, and schedule a time for the harvest. Let your pickers know when and where to show up. Schedule picking to start after the hops have dried from the dew. Do not pick hops when they are wet. If rain interferes you will simply have to reschedule.

Depending on how many varieties of hops you have and when they become mature, you may have to schedule more than one picking date. On picking days have enough chairs on hand for everyone to sit and enough containers for them to put the picked cones into. In our experience people sit around and pick into a single container rather than each person having her own container. This is fine as long as they are all picking the same variety, and somehow it makes the work seem to go faster. However, some individuals really like to see the sum total of their work and will prefer picking into their own container. Containers should be plastic or metal rather than cloth or wicker—something that can be thoroughly cleaned. That being said, it is also best to have a container with mesh sides that will allow for air circulation, like a milk crate lined with screen or a plastic laundry basket. It is amazing how much heat a container of freshly harvested hops generates. It is not as if they are going to burst into flames right inside the picking tub, but high heat such as this degrades the quality of the hops, so the less of it the better.

Remember, too, that you need to have your drying station set up ahead of time. For a small, hand-harvested crop several door or window screens on crates or sawhorses along with a few box fans set up inside a barn or garage will suffice.

CUTTING THE BINE

Since the hop bine is suspended from the trellis by a string of coir, you are going to have to remove the bine from the trellis at harvest time. Before cutting the hop from the trellis, sever the bine 3 feet (0.9 meter) above ground level. Do not cut the bine off at its base, where the coir is pinned to the ground with a W clip. Leaving 3 feet of living bine still attached to

Hops Can Be Toxic to Dogs

As more people begin to brew beer at home, it has been found that hops can be toxic to dogs if they eat them. Both raw hops and hops that are spent after brewing have proven to be toxic. Any breed of dog can be affected, but those that are particularly vulnerable are those breeds, such as greyhounds and border collies, subject to malignant hyperthermia, a dangerously high body temperature with an unknown cause. The major symptom of hop poisoning is a sudden high fever. It is important to cool the dog and seek medical help immediately, as often affected dogs do not survive. It is unclear what exact element in the hop is poisonous to dogs.

Dogs are pretty much a constant presence in our hop yard, and dogs and hops have coexisted on our farm for decades. We have never seen a dog show any interest in eating a hop cone. Perhaps it is the bitter smell that indicates to the dogs the hops are inedible. It seems a dog would be more likely to eat spent hops that have recently come out of the brew kettle and are coated with the sugary malt liquid.

Dogs are ever present in our hop yard and have never shown any interest in eating a hop, but hops can be toxic to dogs.

Woody cuts down bines in the pilot hop yard using a pole pruner.

the roots is better for the plant, as it allows the roots to continue to build up their reserves.

Once the lower end of the bine is cut level you need to cut the coir supporting the bine from the trellis. In some cases the top of the trellis can be reached by climbing a ladder, or you may have some kind of elevated platform available to you. If you can reach the top of the trellis, simply cut the coir with a knife, and the bine will drop to the ground. If you can't get up that high or prefer not to, you can use a pruning pole, like that used for cutting the high branches of trees from ground level. Reach up with the pruning pole and sever the coir at the top of the trellis. Depending on the number of hops and pickers you have, you may want to assign one or two individuals to the task of cutting the bines and bringing them to the pickers, to ensure a continuous flow of work.

Once a bine is cut, it can be dragged to the picking station, where your pickers will remove the cones from the bine and put them into the containers you have provided. Depending on your number of bines and pickers and how heavily the bines are loaded with cones, you can allocate a whole bine to a single picker, pair up pickers on a single bine, or cut a bine into segments and have each picker work on a section. Although you can rig up any number of systems, we have found our volunteer pickers prefer to sit in chairs clustered together so they can chat away while picking cones from bines on their laps, dropping the cones into a container between their feet. Remember to tell the pickers to wear long pants and sleeves as the bines are covered with prickers, similar to nettles. When a bine is dragged across bare legs it can leave welts, and in some cases a bine can cause quite an allergic reaction when coming in contact with skin. Unfortunately it is impractical to wear gloves when picking hops so a picker's hands need to remain unprotected.

The cones are attached to the hop bine by thin stems, which are easy to break, and picking the cones is simply a matter of plucking them from the bine. Of course, pickers have to be careful not to

Erin has sensitive skin so she fashioned these hop-picking sleeves from the legs of a pair of jeans her daughter outgrew.

pull too hard and break the cones apart during this process; but if the hops are not overly mature, the cones should be fairly resilient and remain intact during harvesting.

One great thing about handpicking hops is that it is quite easy to keep debris out of the harvest—unlike in mechanical harvesting, where some plant matter near the cones can get grabbed along with the cones. Make sure to instruct your volunteers to only put hop cones into their picking containers, not bits of leaf, stem, and bine. If debris isn't filtered out at harvest it is just going to have to be done later, as it is unacceptable to have plant matter mixed in with the cones. Not only will plant matter add excess weight to the cones when packaged, chopped-up hop leaves and bines are simply not an ingredient in beer, and no brewer is going to want to use hops that contain such debris.

It is important to keep the cones from different varieties of hops from mixing during harvest.

As the grower your job during the handpicking, besides making sure the beer is flowing, will likely be hustling picked hops to the drying station. One important thing you will have to pay attention to is keeping varieties separate. Large hop growers normally only pick one hop variety at a time, but smaller growers using handpickers can often find themselves harvesting more than one variety on a given day. Granted, many hop varieties have their own unique look. But while you might be able to distinguish the cones of one variety from another, your volunteers certainly will not have this ability. It is easy to get the hop varieties mixed up during picking and drying and highly undesirable to do so. If you don't keep the varieties separate and labeled, you won't know what kind of hops you are offering (or even if they are for brewing or bittering) or how to price them. Getting your varieties mixed up will leave you with a generic product that will be dramatically reduced in both monetary value and usefulness.

Hop bines entering the Dauenhauer harvester on Crosby Hop Farm, which can process an average of 1,000 hop bines an hour.

Mechanical Harvesting

Once you get over, say, 1/4 acre (0.1 hectare) of hops, handpicking is really no longer a viable option. Mechanical hop-harvesting machines range from the very large hop-picking machines and hop combines used in the Northwest to midsize models, such as various Wolf harvesters, to some newer small models being developed especially for the small grower. No matter what the size of the hop-harvesting machine, the basic concept is the same—interfacing rows of mechanical fingers that essentially rake the hop bine, pulling loose the cones.

The capacity of a hop-harvesting machine is generally measured in the number of bines per hour the machine can harvest. This then translates into the hop yard acreage. For example, the Dauenhauer harvester used on Crosby Hop Farm can process an average of 1,000 hop bines an hour. On the opposite end of the spectrum a Wolf WHE 140 harvester, used at Morrisville State College in New York, can process between 120 and 140 bines an hour and is suited for hop farms up to 12 acres (4.9 hectares) in size. The Hop Harvester 1000 that we purchased from Steenland Manufacturing in 2014 is advertised as processing 2 bines a minute, or approximately the same as the Wolf WHE 140, but we found this was not the case. It also did considerable damage to the cones. Steenland is currently in the process of redesigning the machine.

A hop-harvesting machine in Yakima that both cuts the bine from the trellis and begins processing of the bine in the field.

This all makes sense, but the kicker is that these hop harvesters are really, really expensive. How expensive? Well, a harvesting machine on a large commercial hop farm such as Crosby Hop Farm can cost around $1 million. A small Wolf harvester goes for about $30,000. The Hop Harvester 1000 we purchased cost $12,000, which is certainly more affordable, but it didn't work. Another problem with hop-harvesting machines for growers in the East is that these machines are really big and heavy and are only manufactured in faraway places such as California and Europe, which makes it both logistically difficult and expensive to get them to hop farms in the eastern United States. This all adds up to a lot of money for a hop farmer who is just getting started.

Some would-be hop growing regions in the East, such as Madison County, New York, have taken a regional approach, buying one machine and putting it in a central location (a Wolf WHE 140 at Morrisville State College), with farmers cutting their bines,

The Wolf WHE 140 harvester, used at Morrisville State College in New York, which can process between 120 and 140 bines an hour, is well suited to small hop farms in the East and can harvest hops efficiently from hop yards up to 12 acres in size.

trucking them to the machine, then trucking the cones back to the farm for drying. This system can work well, providing the hop farms are near enough to the processor to make it practical, but it takes advance planning because in general everyone's hops are ready at the same time, and the machine can only do so many bines an hour. This is made even trickier by the fact that the cones must be removed as soon as possible after the bines are cut. Others have constructed mobile harvesters that can move from farm to farm. This solves the problem of transporting cut bines to a central harvester, but the fact remains that varieties growing on all the hop yards in a given region will likely be ready to harvest at the same time, so tight scheduling and management of logistics will be necessary. It is ultimately the lack of harvesting infrastructure that is the most significant limiting factor for hop production in the East.

Build Your Own Hop Harvester

Are you handy? I mean really handy? If so, you may want to get your hands dirty and start building your own harvester, or start innovating along with others who are trying to address the need for shared harvesters.

Seeing the lack of hop-harvesting infrastructure facing regional hop farmers, UVM Extension faculty and staff collaborated with hop growers, brewers, engineers, and fabricators to build a functional prototype of a mobile hop-harvesting machine and made the design documents public in a wiki format (see Resources).

The mobile hop harvester is a trailer that can be transported from hop yard to hop yard, on the open road, pulled by a truck with a standard hitch. It runs on hydraulic power off a tractor with a power take-off. Two trained operators working this machine can harvest an acre of hops in a normal workday. The hop bines must be cut and brought to the machine. A bine is then attached to a hook and drawn through a stripping section lined with metal fingers. The cones and vegetation drop onto what is called a dribble belt, which transports the vegetative matter upward. The leaves are carried away and the heavier cones roll back down the belt where they are taken away on their own conveyor belt to a collection point. The machine cost $20,600 in materials and $32,000 in fabricator labor to build.

The mobile hop harvester designed and built by UVM. The design for this harvester is available from UVM in wiki format. Photograph courtesy of Chris Callahan, UVM Extension

Larry Fisher of Foothill Hops, located in Munnsville, New York, also created his own machine. He and a friend, who is a retired machinist, reviewed expired hop-picking machine patents, some from as long ago as 1840, then designed and built a vertical harvester. Not mobile, this harvester is permanently stationed at Foothill Hops. The bines are brought to the machine. A person stands on an elevated platform and attaches the bine to a clip. The bine is moved through two panels of picking fingers, which pull the cones and leaves from it. The leaves and cones drop into a bin, where they are separated. There fans blow the leaves against screens, which transport them away while the cones drop down onto a tray below. It cost Fisher about $12,000 in materials for both the machine and the separator.

Hop harvester designed and built by Larry Fisher of Foothill Hops in Munnsville, New York. Photograph by Michelle Gabel/The *Post Standard*

CHAPTER 12

Drying Hops

Once plucked from the bine, the brew kettle is the ultimate destination for the vast majority of hop cones. What happens to the hops along the way depends on whose brew kettle they are headed for. Brewers use hops in various forms. Some brewers like to make a seasonal beer with freshly picked wet hops—also called fresh hops or green hops—that have not been dried or processed in any way. But all the other forms—dry hops, hop pellets, or hop extracts—require drying the cones first to preserve them. If fresh-picked hops are not used or dried immediately after harvest, they will quickly rot, and the lupulin, along with the alpha and beta acids, will be destroyed.

For as long as people have been growing hops, they have been processing them by drying them. Whether that drying takes place in massive industrial kilns, on ad hoc drying tables made from propped-up screen doors in a living room filled with fans, or in one of the historical oast houses still dotting the rural

Brewer Sam Richardson presides over the brew kettle at Other Half Brewing Company in Brooklyn, New York.

landscape, it is an urgent and delicate process. The quality of the hops you worked hard to grow hangs in the balance: dry them too fast, at too hot a temperature or not enough—and all is lost.

Once hops are dried, they become relatively shelf stable—though eventually the alpha and beta acids and hop oils contained in dried hops will begin to degrade in quality. Improper drying or storage can accelerate that process, but over the years technology has brought new advances.

When you dry hops you are essentially removing most of the moisture from the hop cone. During the process 1 pound (0.5 kilogram) of destined-to-be dry hops must shed 3 pounds (1.4 kilograms) of water, the equivalent of a little bit less than ½ gallon (1.9 liters). Removing the moisture slows the decomposition process of the plant matter and preserves the important alpha and beta acids and hop oil contained in the lupulin. Creating the optimal conditions for drying hops involves an exacting balance of heat and air circulation over a carefully managed period of time. On a commercial hop farm, the low, meditative hum of the drying kiln manned by one or two people is a stark contrast to the bustling human activity surrounding the roar and clatter of the harvesting machine.

How the Large Commercial Growers Do It

On a large commercial hop farm, the harvest is somewhat like a relentless assembly line that runs twenty-four hours a day. If all goes according to plan, hop cones will be dried and baled between twenty-four and thirty-six hours from the time the bine they were growing on was cut. The pace is driven by the necessity of preserving the lupulin contained within

Harvested hops spilling onto the kiln floor.

the hop cones before it starts to degrade. Remember, the higher the level of acids and essential oil in the hop, the higher the price. Field workers move machinery up and down the rows of the hop yard, cutting bines and trucking them to the harvesting machine, where the cones are removed. The harvested cones flow steadily into the drying kilns.

When the cones arrive at the drying kiln they have a moisture content of 76 to 80 percent. In the kilns they are dried to a moisture level between 8 and 10 percent. The harvest can progress only as fast as the hops can be dried. If drying space is limited, which it often is because kilns are extremely expensive to install, the kilning process can be a bottleneck, slowing the harvest. The quality of mature hop cones awaiting harvest will begin to degrade on the bine. The solution to this capacity problem has been to speed up the drying process by increasing the temperature. But that approach can backfire because a higher temperature can degrade the quality of the acids and oils, reducing the value of the hops.

In a commodity market where hops are being purchased cheaply for mass-produced beer, this is not as much of an issue. But lately, as craft brewers become more and more particular about the quality of their hops, growers are lowering the temperatures in their kilns and instead increasing the volume of air circulating or simply giving the hops more time to dry. Historically, hops have been dried at a temperature of about 160 degrees Fahrenheit (71.1 degrees Celsius) over a period of approximately six hours. Today the trend is to dry the cones at lower temperatures over a longer period of time. When we visited Crosby Hop Farm, they were drying their hops at a temperature of 128 degrees Fahrenheit (53.3 degrees Celsius) over an eight-hour period.

In the 1800s, hops were dried on the second floor of barns called hop houses or hop kilns, with rising heat generated by woodstoves located below the drying floor on the first story. This basic concept has remained unchanged. Today a hop-drying kiln on a commercial hop farm is often still a two-story building. Propane boilers located on the lower level blast heat upward through the floor of the second story where the hops lie. The hops are fed onto the floor by the conveyor belt running from the building housing the hop-harvesting machine. The floor is covered with a layer of burlap and divided into sections by short walls about 3 feet (0.9 meter) high. A section will be filled to a depth of about 2 feet (0.6 meter). The flow of hops onto the drying floor is monitored by one employee while another closely tracks the temperature and moisture levels. As the heat and air circulate through the hops, they release their moisture. Leaning over the kilning floor you can feel and smell the warm bitter heat rising against your face. Moisture levels are tested by employees wielding probes who walk out onto the hops wearing large flat boards strapped to their feet, referred to as hop snowshoes, so as not to crush the cones.

Once the hops are sufficiently dry, the burlap underneath them is rolled up—feeding the hops on to another conveyor belt which takes them into the conditioning room. The conditioning room is essentially a big empty warehouse divided into sections. The hop cones are dropped onto the floor in an enormous pile and left to sit for about twenty-four hours. Conditioning is necessary because in the hop

A pair of plywood hop snowshoes lies ready at Goschie Farms. A worker monitoring the drying process will don the snowshoes to walk along the surface of the hops to the center of the kilns to test the moisture level.

Hops heaped high on the conditioning room floor where the temperature and moisture levels are allowed to equalize over a period of about twenty-four hours.

kiln the hops on the bottom, closer to the heat source, become drier than the hops on top. While the hops are in the conditioning room, the moisture levels equalize as the temperature drops. This process is also called cooling or curing. Once they are done conditioning, the hops are baled.

Drying Options for Home Growers

Obviously a hobby hop grower is not going to need all this big expensive equipment, but the basic principles remain the same. After your hops are picked off the bine they are very moist. If they are not used immediately in a wet-hopped brew, they will begin to rot. How you dry them really depends on how many hops you have.

Air drying can work well, providing you have enough space. One method of air drying is to spread the hops out on a horizontal screen. Make sure the screen is clean and dry before you put the hops on it. Elevate the screen like a tabletop. That way you are getting air on the hops from both above and below. Enhance air circulation by setting up fans around the screen. Make sure the hops are out of the sun, as sun drying will compromise their quality. How long it will take to dry hops using this method will really depend on the weather. In warm, dry weather it could only take a couple of days. In wet, humid weather it could take a week. Check them several times a day, and stir and turn them a bit.

The Historical Oast House

In their early days of cultivation, hops were generally dried in the open air, but when commercial hop production ramped up, economic viability demanded the drying process speed up. Given the resources available at the time, the best way to achieve this was by applying heat—and large kilns called oast houses began popping up in England and other beer-brewing regions of Europe as early as the sixteenth century. An early oast house was essentially a one-story building with curved walls and a wood-fired brick furnace located on the ground. A wooden slat floor was installed a few feet above the furnace, with about ¼ inch (0.6 centimeters) of space in between each slat. The hops were spread on this floor. Heat rose up through the slatted floor from the furnace below, and the heat and moisture were vented through a roof shaped like an upside down ice cream cone.

Oast houses based on this design were built in the English colonies, too, and over time innovations were made. In the 1800s a typical oast house, by then called a hop kiln, was a vertical two-story building, usually built from stone into a bank or hill with a pyramid-shaped roof. A coal-fired furnace on the bottom floor generated the heat, which rose up through the slatted floor and vented through the roof. The drying process was also referred to at this time as "curing." As hop production increased, the hop kilns were added onto, creating four or even more bays, each with its own furnace. In many cases the dried hops were stored in the same building that housed the kiln. As you can imagine, fires were not unheard of and proved disastrous when not only the kiln but also the entire processed and stored crop was lost.

Regulating temperature was tough under these conditions, and hops were often overdried, in some cases turning brown during the process. Sulfur was burned in the kiln to green the hops up again after drying but even though the color changed, the damage was done. Many diatribes were written in newspapers and hop industry publications about the problem of ruining hops with high heat. In 1883 Ezra Meeker, author of *Hop Culture in the United States*, wrote, "Almost any inexperienced farmer can raise hops, but nothing short of the most vigilant, careful and intelligent management will prepare the crop, without injury, ready for market; hence the curing is the all important part of hop growing and if not properly done, results in great loss and final failure."

An old-style hop kiln from the 1800s still stands in Clarksville, New York.

In this picture from our early days growing hops for home use, we are drying hops on a screen. The chicken wire that can be seen lying over the top of the hops has nothing to do with the drying process but is instead being used to deter a curious barn cat.

Clever people have come up with a lot of different ways to dry hops at home and shared them online, ranging from outfitting a cabinet with a hair dryer to sandwiching hops between furnace filters and strapping them to a box fan with a bungee cord. But more familiar kitchen methods work, too. A food dehydrator can be used to dry hops, with the major limiting factor being its low capacity. This method works well for home brewers who are harvesting a pound or so of wet hops a day and want to put them in the dehydrator overnight. Be careful to set the dehydrator between 120 and 140 degrees Fahrenheit (48.9 and 60 degrees Celsius), and dry the hops overnight. Hops can also be dried in the oven on the warm setting providing it does not reach over 140 degrees Fahrenheit. Set your oven on warm, and let it run for a while, then put a thermometer in it to see what the temperature is before putting in your hops. Small amounts of hops can be dried in the microwave in a matter of minutes, but this is not recommended because the temperature will get too high and damage the hops.

Drying Options for Small-Scale Commercial Growers

Although some manufacturers have begun to produce hop-harvesting machinery for small-scale commercial hop farms, there is currently a real missing link in between the box fan and screen method and the technologically advanced hop kilns in the Northwest. Small-scale growers have the option of building their own oast houses. If you have an engineering bent you can devise your own system—or you can download plans online.

The modular oast house designed by UVM has numerous drawers, enabling multiple varieties of hops to be dried at the same time without mixing. Plans are available online.

This drying kiln built by Steenland Manufacturing uses only air circulation to dry the hops and works very well.

UVM has designed and made plans available for a modular oast house that is essentially a heated cabinet outfitted with screened trays in the form of drawers and fans (see Resources). The system of drawers is a great way to keep hop varieties separate during drying. It is 4 feet (1.2 meters) wide, 4 feet deep, and 8 feet (2.4 meters) tall and made with readily available materials. This unit can bring 300 pounds (136.1 kilograms) of wet hops down to a moisture level of 10 percent in eight hours.

In the past we have relied on a drying machine from Steenland to dry all of our hops. Its limited capacity slowed down our harvest to a degree, but it worked out. Prior to that we relied exclusively on the box fan and screen method, except in the case of the collapsed trellis, in which the hops dried on the living room floor. Now that we have grown, we are purchasing a hop dryer manufactured by Wolf (for $4,800) with multiple layers and enough drying capacity to process up to 30 acres (12.1 hectares) of hops. Although we will not need all of this drying capacity for our own hops, we plan to offer drying services to other growers in our region since so many are suffering from a lack of infrastructure. One advantage of getting started growing hops is that they take several years to mature and reach full production. Your first harvests will be small, giving you time to experiment and ramp up to the purchase of infrastructure that will allow you to process your crop once the hop yard comes into full production.

How to Tell When Hops Are Dry

When hops are first picked, their moisture level should be between 76 and 80 percent of their weight. Hops are considered dry when the moisture level has been reduced to between 8 and 10 percent of their weight. When the hops are dry they will feel dry even when you squeeze them. When the cone is dry the stems should snap, not bend under pressure. To make a scientific calculation, use the same formula that you used to decide when to harvest.

Large farms use moisture probes to test moisture levels. The problem is these are larger than what small-scale growers need and are prohibitively expensive. When we visited Goschie Farms, they were using a moisture meter called the Moist-VU DL6000 made by Reid Instruments. It costs over

$7,000. That prompted us to call Tom Reid of Reid Instruments and see if there were options for a farm our size. It turned out that he had been getting a lot of demand from marijuana growers for a smaller moisture meter, so he had been working on developing a tabletop version called the DL4000 that he thinks would work for small-scale hop growers as well. He estimates it will cost about $4,500. That seems like a lot of money, but when you consider that the success of the whole crop depends on the accuracy of the moisture reading, it is worth it.

Since our initial phone call with Tom Reid, Reid Instruments is selling the Moist-Vu DL4000 (see Resources). It measures the moisture level of a bagged sample of loose cones placed on a sensor plate using the same ultrasonic measuring device as the moisture probe. Reid Instruments is providing us with a model for trial use this fall.

There is a lot at stake in getting the dryness/moisture ratio right. If you dry your cones to a lower moisture percentage, they will begin to fall apart and the lupulin inside will disintegrate and be lost. If the hops are too moist, you run the risk of their oxidizing, turning brown and moldy, or even heating up and starting a fire through spontaneous combustion—like hay that is still wet when baled. Normally all hop bales going into storage are tested with a moisture probe designed for bales to make sure they have been properly dried. But 4 percent of the nation's total hop crop in October of 2006 was lost as 2 million pounds (907,000 kilograms) of hops burned in a Yakima warehouse fire resulting from the spontaneous combustion of hops not fully dried. "Yeah, that was us," said a sheepish Paul Matthews, of Hopsteiner, who told us the story.

The moisture probe used on Goschie Farms, the Moist-VU DL6000 made by Reid Instruments, is used to determine the moisture level of a layer of hops lying 2 feet (0.6 meter) deep on the kiln floor.

CHAPTER 13

Preparing Hops for Use and Sale: Analyzing, Packaging, and Pricing

Whether you are growing hops for your own brew pot or someone else's, processing hops has many steps—from harvesting and drying to packaging and storing. If you're selling to others, you'll also need to analyze your crop and make decisions about how to price it. Many of your decisions along the way will be driven by the form of hops you or the brewers you are selling to want—wet, dried, pelletized, or a hop extract.

In the past, not many buyers have wanted wet hops—also called fresh hops or green hops—because they have to be used immediately. But interest in them is on the rise. And for growers who want the simplest way to sell small quantities, this is good news. Here in the East, where locally grown hops are scarce, a brewer at a local brewpub might be interested in buying them simply because they are grown nearby. Other brewers are simply eager to experiment with wet hops, since they haven't had ready access to them in the past. And some brewers are so excited about wet hops that they come straight to the hop yard with their coworkers to pick the hops themselves and bring them back to the brewery. Because of their fresh nature, wet hops are not packaged but instead are transported directly to the brewery in bags, baskets, or boxes that remain open so that the hops can breathe. Once at the brewery they are refrigerated, then quickly used.

Most hops even small-scale growers sell, though, will be dry. And once the hops are dry it is critically important that they are analyzed and packaged quickly.

Hop Analysis

It is important to remember that it is not the hop cone itself that the brewer is after but the alpha and beta acids and hop oil contained within the lupulin. So one of the first steps in preparing your hops to go to market is to find out exactly what's in that lupulin you worked so hard to produce. To do that, growers send hops to laboratories for analysis to obtain information brewers will want prior to purchasing the hops and using them to brew beer. Hop analysis is conducted in laboratories using scientific methods that have been standardized by the American Society of Brewing Chemists. Depending on the types of tests you choose to have done, the analysis can indicate dry matter content, information on the levels of alpha and beta acids and oil content, and a profile of

The Brewer's Cut

When brewers are evaluating hops before purchase they consider the hop analysis report from the laboratory, but an in-person sensory evaluation also takes place. A brewer's cut is essentially a core sample from a 200-pound (91-kilogram) bale of dried hops that is given to the brewery for examination. Hop sellers often provide a special room for this evaluation where hop buyers evaluate many different samples. They rub the hop cones between their hands raising a cloud of hop fragments and lupulin dust. During this process they are evaluating the aroma of the hops, the stickiness of the lupulin, level of dryness, and many other elements they may seem intangible to the lay person.

The brewer's cut is a small segment taken from the center of a 200-pound (91-kilogram) bale of hops that allows brewers to conduct a sensory analysis of the quality of the hops to help them decide what to buy.

the chemical compounds contained within the volatile oil. It can also provide information about the hop storage index, which is essentially a prediction of the pace at which the alpha and beta acids and oils in the hop will deteriorate through oxidation over time. Large commercial hop producers also conduct analysis in the field before harvest to help determine when alpha and beta acids are optimum and the hops are ready to be picked.

There are several laboratories that analyze hops. Based on the type of test and the laboratory's requirements, the sample submitted can be anywhere from 1 to 8 ounces (28 to 227 grams) from a given lot of hops and cost between $25 and $100. Although fresh-picked hops can be tested, they don't hold up well in shipping. Hops submitted for testing must be dried to a level of 8 to 10 percent moisture. For best results, submit frozen dry hops. Some labs provide sample kits; otherwise the sample of hops should be put in a ziplock bag and boxed with a freezer pack, then mailed overnight to the lab. The analysis will usually be received by the grower within a couple of days to a week's time.

If you are a home brewer, the analysis of your hops will help you understand how to best brew with them. If you are planning on selling your hops, your marketing plan will depend on the results of the analysis. Brewers will review the analysis along with the look, feel, and smell of your hops to determine if the hops are suited to the type of beer they want to produce.

HOP ANALYSIS LABORATORIES

In the East, hop-quality analysis is performed at the University of Vermont Hop Quality Testing Laboratory, which operates under the umbrella of the University of Vermont Cereal Grain Testing Lab. The lab will provide you with your hop sample's brewing values, which include the levels of alpha and beta acids as well as the hop storage index (see Resources).

Alpha Analytics is a laboratory in Yakima, Washington, that offers analysis of dry matter and brewing

203 Division Street Yakima, WA 98902
Phone: 1-877-875-6642 | Fax: 1-509-895-7968
E-mail: info@alphaanalyticstesting.com

Certificate of Analysis

Date Issued: October 3, 2014
Analysis Results Issued To: Steve Miller -- Cornell Cooperative Extension, Madison Cty

Brewing Values

Sample ID	Analysis Date	Variety	Description	Alpha Acid %	Beta Acid %	HSI	Oil %
H140292	**10/3/14**	Cascade	dietrich gehring	5.6	7.3	0.211	1.08
H140293	**10/3/14**	Nugget	dietrich 2	11.6	4.3	0.237	1.26
H140294	**10/3/14**	Columbus	dietrich	10.3	6.7	0.225	2.10
H140295	**10/3/14**	Cluster	dietrich 3	9.8	6.8	0.207	0.45
H140296	**10/3/14**	Brewer's Gold	dietrich 4	7.4	5.3	0.233	1.86
H140297	**10/3/14**	Centennial	dietrich 5	9.5	3.9	0.244	2.33

Notes:
NO GROWER INFO PROVIDED

Method of Analysis

ASBC Hops 6a – α-acids/β-acids by Spectrophotometry
ASBC Hops 12 – Hop Storage Index (H.S.I.)
ASBC Hops 13 – Total Oil by Distillation

Issued By

Zac German
Alpha Analytics Laboratory Manager

This report from Alpha Analytics shows the test results for several varieties of hops from our 2014 crop.

values as well as hop oil. The laboratory can provide you with the content of hop oil simply as a percentage and also offers a volatile oil profile, which tells you what compounds are in the oil at what levels (see Resources).

Packaging Small Quantities

The point of swift packaging is to slow down the oxidation of the hop cones by limiting their contact with oxygen. If you are a small grower and can't package your dried hop cones right away, it is best to freeze or refrigerate them until you can package them. This will slow down the oxidation process. Otherwise, packaging should start as soon as the hops are dried and cured. The range of packaging options varies with the quantity of hops you are trying to deal with. It makes sense, though, to package the hops in the quantity that the brewer you are selling the hops to will use them in.

Dry whole hop cones are referred to as "leaf" in industry parlance. Large producers generally compress dry hops into 200-pound bales. A 200-pound bale priced by the pound is considered the most economical way to purchase hops. This is also the size generally loaded into hop pelletizers. But really, a bale can be any size as long as its weight is known, as it will ultimately be priced by the pound. For example, Crosby Hop Farm starts out with 200-pound bales, which they pelletize. If the hops are not to be pelletized but instead to be sold as leaf, the 200-pound bales are cut into four sections weighing 50 pounds (22.7 kilograms) each and vacuum sealed. These 50-pound packages are purchased by craft brewers who want leaf; 10-pound (4.5-kilogram) packages of hops are also available.

But brewing on a smaller scale calls for smaller packages. If you are packaging for home brewers, break the hops up into quantities of 2 ounces (56.7 grams). If the brewer is working on a somewhat larger scale, create 1-pound (0.5-kilogram) packages. You will need to invest in an appropriate-size digital scale to measure the desired amounts. By providing the hops in usable amounts, you are avoiding a situation in which the brewers are going to have to open the same package over and over again to extract the amount of hops they need for a given recipe. This repeated opening and closing of the package will increase the oxidation of the hops.

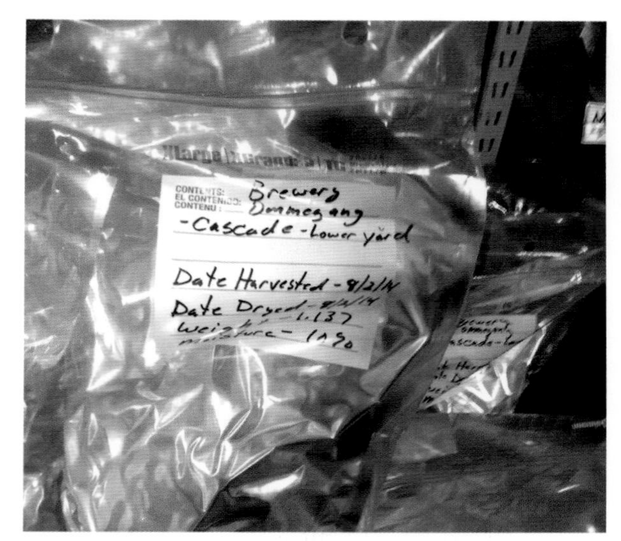

Smaller sealed bags of pelletized hops from the hop yard at Brewery Ommegang in Cooperstown, New York.

For smaller packages, sucking the oxygen out of home freezer bags can work. But the best method is to vacuum seal the hops in Mylar oxygen barrier bags, which you can do with either a home use vacuum-sealing machine or commercial vacuum-sealing equipment—though once you are producing more hops, you may want to invest in commercial equipment. Essentially the oxygen must be sucked out of the bag; then the bag is sealed airtight. For even better results, some hop processors flush the bag with nitrogen before vacuum sealing, which requires a commercial sealer as well as nitrogen flushing equipment. We purchased a FoodSaver sealer for about $150, and that has been okay for the time being, but we are going to have to upgrade. A commercial vacuum sealer can cost $2,000 or more.

Whether you are using a home vacuum sealer or a high-tech commercial system, check to see whether your state requires you to have a food-processing license to vacuum seal an agricultural product for sale. New York, for instance, requires an Article 20-C Food Processing Establishment License from the state. And remember to label your packages with the name of the variety, weight of the package, date, and the results from the hop analysis.

Once you have your hops sealed into airtight packages, put them in a freezer for the best results; second best is a refrigerator.

Baling Hops

If you plan to sell hops in larger quantities, you will need to bale them. If your hop yard is small, it is probably extremely premature to think about 200-pound bales; however, a smaller bale is still a good way to handle larger amounts of hops, and once

A Prius loaded with our minibales of hops on the way to the hop pelletizer. Photograph by Laura Ten Eyck

Six bales of hops weighing 200 pounds each (90.7-kilograms) in the warehouse at Crosby Hop Farm.

Homemade Hop Balers

This hop baler made from a log splitter by UVM produces 20-pound (9.1-kilogram) bales.

When it comes to making a 200-pound (91-kilogram) bale of hops, you obviously need special equipment. But if you are a small producer you can make smaller bales with improvised equipment. Using off-the-shelf materials, engineering students at UVM adapted a log splitter to make 20-pound hop bales for under $1,500 (see Resources).

We took Foothill Hops' lead and converted a trash compactor into a hop baler. After attempting unsuccessfully to purchase a used trash compactor, we bought a new Kenmore at Sears. Since the pressure of a household trash compactor is not sufficient enough to create a dense bale of hops, we cut wooden spacers the same dimension as the interior of the compactor and ½ inch thick. We kept adding them and experimenting with compaction density. We ended up stacking five spacers on the bottom and placing one on top to create a small bale weighing about 6½ pounds (3 kilograms). We did not have equipment to vacuum seal a bale this size, so we ended up simply packaging them in plastic wrap. Because of either the nonairtight packaging or insufficiently dried hops, we did have some issues with moisture prior to pelletizing, so we still have a way to go in terms of technology.

Hops going into our trash compactor baler to create a minibale weighing about 6½ pounds.

again growers have proved ingenious in their designs. The folks at UVM's research farm converted a log splitter into a hop baler; another scheme involves a trash compactor (see sidebar). Once you have your bales, wrap and seal them in plastic and store them in the freezer. If you are going to have your hops pelletized, bales are a handy way to deliver them to the pelletizer.

Keep storage needs in mind. As a hop grower you must have enough space to store your product at

Vacuum-sealed bags of hop pellets in the cooler at the Northern Eagle Hop Pelletizing Company in Oneonta, New York.

various stages during processing. This means you will need to have enough storage capacity for the bales of dried hops before they are sold or brought to the pelletizer. From 1 acre (0.4 hectare) of mature hops you will produce between 1,000 and 2,000 pounds (454 to 908 kilograms) of hops (between five and ten 200-pound bales). The storage space must be refrigerated with controlled humidity.

Hop Pelletizing

It is very common for hop cones, once dried, to be processed into pellet form through a combination of heat and pressure. This not only extends the product's shelf life but also takes up a lot less space, making pellets easier to store as well as ship.

Refrigerated warehouse at Crosby Hop Farm where cases containing vacuum-sealed pelletized hops are stored.

Hop pellets are basically tiny cylinders of compressed hops. If you have ever cared for domestic rabbits or guinea pigs, you will be struck by the apparent similarity between pelletized hops and feed for these animals, which are made primarily of pelletized alfalfa. In fact, the process for pelletizing hops is basically the same as the process for pelletizing animal feed. Once dried, the hop cones are put through a hammer mill, where they are crushed by multiple blows from small metal hammers rotating at high speed. After milling, the hops emerge in a loose flake form. The crushed hops are then fed into a pelletizing machine. This machine includes a die machine, which comprises a metal plate lined with pellet-sized holes and a rolling cylinder. The crushed hops are forced through the holes in the plate by the pressure from the roller. They emerge from the other side of the plate in the form of ¼-inch (0.6-centimeter)-long pellets.

There are essentially two different types of pellets. T-90 pellet hops have simply been turned into pellet form via the above process. T-45 pellets are the product of a more elaborate process in which some of the hop vegetation is removed and the acids are concentrated.

Many craft brewers prefer to use hops in pellet form for several reasons. They are more shelf stable and less prone to oxidation than hop cones and are easier to measure and blend with accuracy. During the boiling phase of brewing, hop pellets take up less space in the brew kettle, don't absorb as much of the wort, and dissolve into small particles—whereas hop cones break down into loose vegetation that can clog up the works. This dissolution into small particles also means the chemical compounds in hops can be more fully extracted from the vegetation during the boil.

Another factor is storage, which can be a problem for brewers, too—especially in the urban areas where so many of them are located. Pellets take up half as much space as cones, although because they are concentrated they also take longer to use—which means the space they do take up they will occupy longer.

Dried hop cones are crushed and compressed into pellets that not only take up a lot less space than whole cones but also have a significantly longer shelf life.

One concern about pelletizing is that the process can alter the chemical compounds in the hops. This is because all that hammering and compression that goes into making hops into pellets generates heat, and as we know, too much heat is bad news for hops. Pelletizers processing hops can reach temperatures of 160 degrees Fahrenheit (71.1 degrees Celsius). If temperatures are allowed to get that high it means trouble because the chemical compounds in hops start to degrade when temperatures reach 100 degrees Fahrenheit (37.8 degrees Celsius).

Some hop pelletizers are designed to operate at cooler temperatures. Hop growers such as Crosby Hop Farm, sensitive to such quality issues, sometimes purchase their own pelletizing equipment so they can control the process. Crosby Hop Farm only runs their pelletizer early in the morning to avoid operating in the heat of the day and risking damage to the hops with too much heat. Other growers take advantage of Crosby's initiative. For example, Goschie Farms has their hops pelletized at Crosby Hop Farm because they know they can trust another hop grower to do the job right. Here in New York, Foothill Hops

East Hop Terroir

Terroir is a French term that refers to flavors imparted to a particular agriculturally produced ingredient that are believed to come from the specific conditions of the locale in which it was grown. Originally a hot topic among wine producers, terroir is now considered an element in the production of many food products ranging from cheeses to meats and now beer. Terroir is attributed to the impact a particular location's features—such as soil, water, temperature, humidity, and day length—have on the genetics of a particular crop. Environmental conditions can influence what's called the "methylation" of DNA, a process whereby certain genes are switched on and others off.

The terroir of beer results from where the beer's ingredients were grown and how the brewer made use of their unique characteristics during the brewing process. A study by Oregon State University demonstrated different essential oil levels in the same hop varieties grown on farms only a few miles apart. Although it is too early to know for sure, it appears that the humid environment of the East may be lending itself to the development of stronger aroma properties (indicated by higher levels of essential oils) in hops grown in this region. Some research has indicated that hops produce more essential oils in their effort to fight off fungal pathogens such as downy mildew. Certainly analysis of hops on our farm have shown higher levels of essential oils than is typical of what's found in those same varieties when grown in the West. In fact, the high oil content is one of the first things that attracted the attention of the brewers who have bought our hops.

purchased a Pellet Pros Model PP220 from Pellet Pros, located in Iowa, that they found operates at a cooler temperature than other machines they experimented with.

Pelletizing machines come in a range of sizes. Production is measured in the number of pounds of raw ingredient a pelletizer can process in an hour. Some hop farms do their own pelletizing, and some send their hops out to be pelletized. Small-scale pelletizing equipment that can process 80 to 100 pounds (36 to 45 kilograms) of hops an hour is available for around $3,000. That does not include a hammer mill, which can be purchased for around $2,000. The more pounds of hops the equipment can process in an hour, the more expensive it is. Small-scale, specialized pelletizing machines designed to minimize heat during processing can cost $40,000-plus, including the hammer mill. Another thing to keep in mind is that the die and roller parts do not last forever. The equipment's life span depends on the materials being pelletized, but in general each die and roller set is good for between 400 and 600 hours of processing. Replacement parts are available, but they are not cheap. For example, a replacement die and roller set for the PP220 is $300.

If you do decide to purchase pelletizing equipment, just be aware that there is more to the job than just turning the hops into pellets. Once the hops are pelletized they have to be vacuum sealed. As mentioned previously, this involves more equipment and (in some states) also involves a food-processing license.

If you don't want to get into the pelletizing business, you will be sending your hops elsewhere to be pelletized. Make sure the pelletizer you use is committed to high quality standards. In addition to ensuring that they control heat during processing, you are going to want to make sure they process your hops as soon after receiving them as possible. You don't want your hops languishing in storage, oxidizing all their good hoppiness away. Also, for the same reason, be sure that your hops are vacuum sealed as soon as they are cooled postpelletizing.

Quality considerations aside, there simply are not very many pelletizers currently in operation in the East. At the time of this writing there are only three pelletizers in the Northeast, and they are all located in New York: Northern Eagle Hop Pelletizing Company, in Oneonta; Whipple Brothers Farms in Kendall; and Foothill Hops in Munnsville, about 40 miles (64.4 kilometers) east of Syracuse. So there is plenty of opportunity for entrepreneurs to address the lack of pelletizing as well as harvesting and drying infrastructure.

Hop Extracts

Kicking it up a notch, hop-extraction plants orchestrate extreme heat, intense pressure, and complex chemical reactions to reduce hop cones into bitter resin and concentrated oil. Craft brewers sometimes use these extracts to escalate bitterness. Large commercial brewers, such as MillerCoors, tend to rely heavily on liquid hop extracts but also still use pelletized hops. The extract from 1 acre (0.4 hectare) of hops will fill a 45-gallon (170.3 liter) drum.

How Hops Are Priced and Sold

Whether you are pricing leaf, pellets, or extract, the price of processed hops is generally based on weight. Because you don't need a lot of hops to brew small batches of beer, hops sold to home brewers often are priced by the ounce, while hops sold to small-scale brewers are generally priced by the pound. As the degree of processing goes up, so does the price per pound. And as with most agricultural products, the larger the volume purchased, the cheaper the price.

Another factor in pricing is whether the hops are being purchased "on contract" or "spot purchased." Brewers often purchase hops on contract one, two, or even several years ahead. Contract purchasing basically means that the brewer is obligated by contract to buy an agreed-upon amount of hops from a grower and the grower is obligated by contract to sell that amount to the brewer. This benefits the grower, who then knows how much to grow in advance and has a guaranteed customer. The brewer benefits from having a secure supply of hops at a set price. Another advantage to the brewer is that hops purchased on contract are cheaper.

The alternative to buying hops on contract is spot purchasing. Hops available for spot purchasing are essentially any hops a grower has produced above and beyond what is needed to fill the grower's existing contracts. Hops spot purchased are more expensive than hops purchased on contract. The advantage to spot purchasing for the brewer is the flexibility it offers. If a brewery that is spot purchasing finds it does not need as many hops as it had anticipated, it can simply buy less. Purchasing on spot also allows brewers to change recipes and try new things on short notice. The advantage of spot purchasing for the grower is first and foremost the higher price, but offering hops for spot purchase also gives you an outlet in the event of overproduction and allows you to provide customers with a level of flexibility and convenience.

Wet hops are priced by the pound. They obviously weigh more than dry hops. In fact, they weigh a lot more; so the price per pound for wet hops is normally much lower than that for dry. This also makes sense when you consider the fact that the grower has less invested in wet hops than dry hops because there has not been the expense of drying and packaging. However, unless you are only growing a small number of hops, it is unlikely that you will be able to sell all of your hops wet.

In general, pelletized hops are vacuum sealed and boxed in 44-pound (20-kilogram) and 11-pound

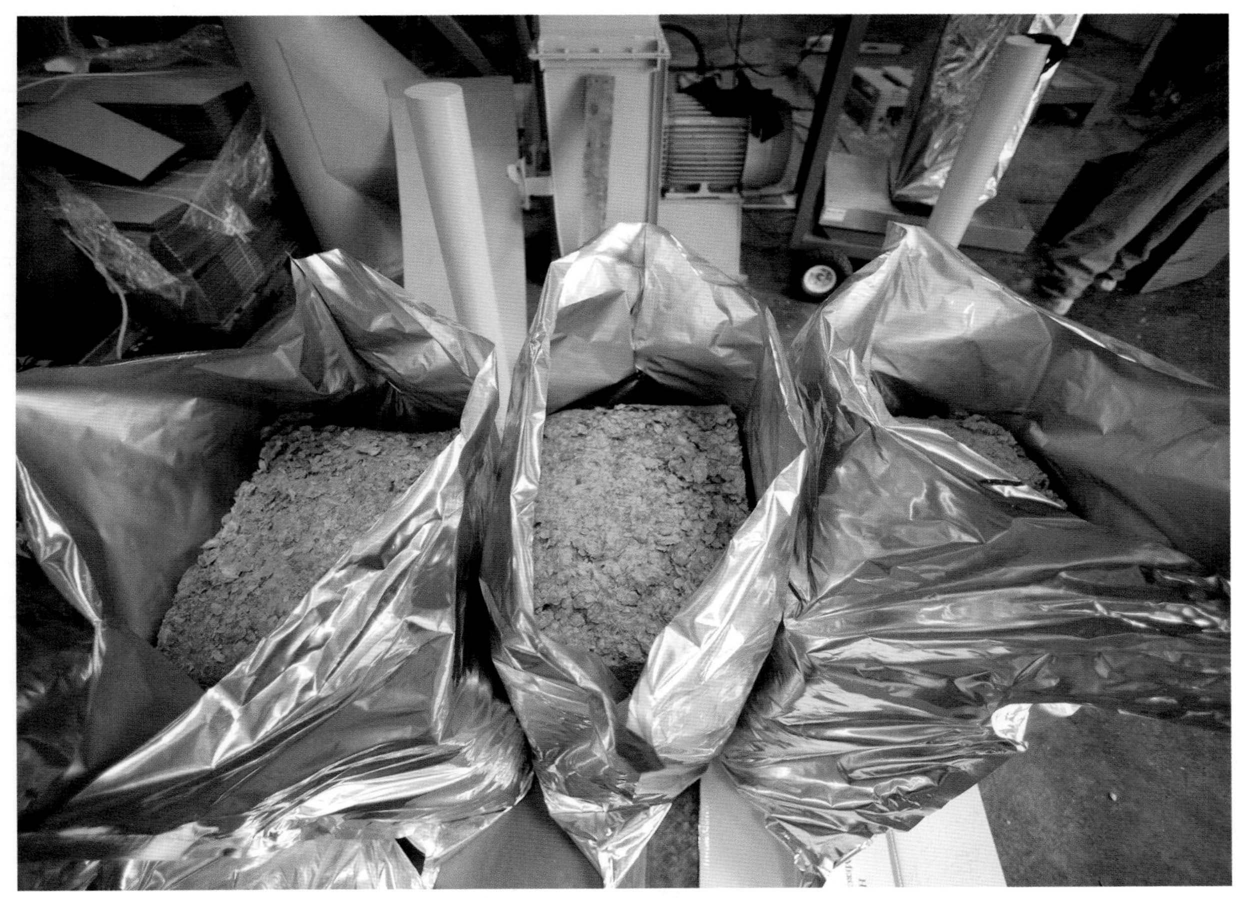
Compressed whole leaf hops in a Mylar bag before sealing at Crosby Hop Farm.

A bushel basket of fresh-picked wet hops in our pilot hop yard.

(5-kilogram) cases. Liquid hop extracts in small amounts come in glass or plastic jars or vials. Larger amounts are packaged in aluminum bottles, cans, pails, kegs, drums, or barrels. Even though extract is liquid, it is still sold by the pound. The price of a pound of hops depends on the volume being purchased, whether the hops are bought on contract or spot purchased, and the degree of processing.

Hop prices also go up and down from season to season depending on the quantity and quality of the crop, both regionally and globally. Within this context different varieties of hops sell for different prices. To give an idea of the range of prices, let's look at the 2014 pricing for spot-purchased hops in the Northwest:

- Leaf packaged in a 200-pound (91-kilogram) bale: about $1,600 a bale on average, which works out to $8 a pound, depending on the variety.
- A 50-pound (22.7-kilogram) package of leaf: between $6 and $13 a pound, depending on the variety.
- A 10-pound (4.5-kilogram) package of leaf: between $7 and $14 a pound, depending on the variety.
- Pelletized hops: slightly more expensive than leaf, with a 44-pound package of Cascade hop pellets at about $12 a pound and an 11-pound package at $12.50 a pound.
- Hop oil concentrate: on the absolute far end of the price spectrum, a single pound can sell for $1,000.

Hop pellets vacuum sealed in various-size Mylar bags.

At this writing, craft brewers in the East have been getting the vast majority of their hops from the Northwest and have been paying prices in the range of what is described above. In addition they are able to choose from a large selection of readily available varieties. In contrast, newly established hop growers in the East are currently producing a limited number of varieties in extremely limited quantities. Because of the learning curve, the quality of the hop cones can also be an issue. Fresh off significant investments in rhizomes, trellis systems, and equipment, growers are anxious to begin recouping their costs, but their ability to set prices is greatly influenced by current pricing charged by the well-established hop yards of the Northwest.

While brewers are enthusiastic about producing beer brewed with locally grown hops, it remains to be seen what price the Eastern market will bear. This will depend in large part on consumer interest. In 2014 Eastern hop growers sold their leaf and pelletized hops directly to brewers for between $6 and $12 a pound, with wet hops going for between $4 and $8 a pound.

A single mature hop bine produces about 1½ pounds (0.7 kilograms) of dried hops. With about 900 bines per acre, a hop yard 1 acre (0.4 hectare) in size could produce about 1,350 pounds (612.4 kilograms) of dried hops. If those were to sell for between $6 and $12 a pound, 1 acre could bring in between $8,100 and $16,200. In 2014 there were only approximately 300 acres (121.4 hectares) of hops in production in New York and New England. The majority of these acres were planted after 2012 and have not yet reached full production. It is too soon to tell what yields will be in the East or what price the market will ultimately pay. But the number of craft brewers and brewpubs in the region continues to grow, and hop growers are striving to meet the demand for local hops.

CHAPTER 14

Brewing Beer

Last year, while at a hop conference, we were standing outside in a snowstorm with Sam Richardson and Matt Monahan of the Other Half Brewing Company, drinking beer poured from growlers they had brought up from their brewery for the occasion. Richardson, the head brewer, is a bearded and sprightly man who hails from Portland, Oregon. He has been brewing beer for many years and is an example of one of the young brewer-slash-mystics that are driving the continued growth of the craft beer industry beyond all expectations. His brewery is tucked away in a corner of Brooklyn underneath a highway overpass and across the street from a McDonald's and a junkyard. Yet just go around the corner, and you will find yourself in the trendy Carroll Gardens neighborhood featuring yoga studios and sidewalk cafes. The beer we tasted that day was a testament to Richardson's brewing talent. I asked

Sam Richardson in the tasting room at the Other Half Brewing Company in Brooklyn.

him if brewing beer was more like chemistry or cooking. He said it was both. To me it seems like alchemy.

Richardson soon sent another brewer our way. Like Richardson, Dan Suarez is small and almost monkish in the simplicity of his ways and mildness of his manner. In our small brewery, he wears a Carhartt knit cap, a red flannel shirt under a hoodie, and rubber shoes. He came to try a recipe for a farmhouse saison of his own invention, using only ingredients from our farm—including malting barley, hops, and blueberries. From a tiny, bell-shaped jar he drank the water as it heated and inhaled the steam from the brew kettle. He sipped the simmering wort, the unfermented liquid that results from the mashing of grains. It was clear his mind was processing a swarm of information based on taste, temperature, and texture. In no time, he was making numerous decisions and adjusting dials, stirring pots, allowing for the passage of time. He called out for ¾ ounce (21.3 grams) of Nugget hops and, when the pellets appeared, dropped them into the boil—a roiling brown liquid with beige foam coagulating on the top. Upon hitting the liquid, the pellets disintegrated, vanished, then rose up in a scattering of tiny green particles that turned the foam olive green. At the end of the boil, he called out again, this time for 7 ounces (198.5 grams) of Brewer's Gold. After the boil the beer was poured off into the fermenter, the yeast was added, and the concoction let alone to ferment. Days later he e-mailed to ask that we add 35 pounds (15.9 kilograms) of blueberries to the fermenter. Two weeks later Dieter gave it a try and said it was amazing. After it spent another three months in the fermenter the beer was bottled.

Dan Suarez samples the wort while brewing a trial batch of blueberry farmhouse saison. Photograph by Laura Ten Eyck

Home, micro, and craft brewers come up with a seemingly endless variety of tastes, but it all starts with four basic ingredients—grain, water, hops, and yeast. There are countless guides to home brewing (see Resources), but it's worth reviewing some basics here, and some recipes in the next chapter, for growers who want to understand what goes into the process.

Grain

Grain is the most substantial component in beer. It gives the beer its color, basic flavor, texture, foam, and carbonation. A variety of grains can be used to make beer, but barley is by far the most commonly used. Barley, a member of the grass family, was one of the first grains to be cultivated and today is the fourth-largest cereal crop produced globally. As with hops, there are different varieties of barley bred for different purposes. These varieties fall into two groups—malting barley and feed barley. As its name implies, feed barley is grown to feed livestock. Over half of the barley grown in the United States is for this purpose. Despite its considerable health benefits, only 2 percent of the barley grown in the United States is destined for human consumption. Malting barley is grown to be malted for the brewing of beer.

A field of malting barley growing on our farm.

Megan, our friend who helps out on the farm during summers off from college, riding the combine during barley harvest.

Malting Barley

The part of the barely used in brewing is the kernels, which are essentially the seeds and the casing that surrounds them. Before it can be used in brewing, barley must be malted, a process that begins by wetting the kernels to force the germination of the seeds inside them. This starts the process of breaking down the natural malt sugars within the seeds' endosperm—the starch inside the seed.

Once the barley has begun to sprout, germination is halted by heating the barley. If the barley is destined to be made into a pale ale or light-colored beer, such as a lager or a pilsner, the process stops there. If it is destined for a darker beer, the heating continues until the malted barley begins to darken to the desired degree. Once this process is accomplished, the barley officially becomes known as malt or barley malt. If needed for a particular beer type, the malted barley can be further roasted through kilning.

As a final step the barley malt must be milled, a process that involves cracking the grain's hard outer husk to provide access to the sugar contained within the kernels. Hot water is then run through the cracked barley malt in a process called mashing because of the porridgelike mash the barley malt becomes when wet. During mashing, the water leaches the sugar from the malt. The spent malt is then removed from the scene—leaving behind rich, brown, sugary water called wort.

Barley tassels showing individual grains before harvest.

Over 40 percent of the barley produced in the United States is made into beer. You can use malting barley to feed to livestock, but you cannot use feed barley to make beer.

Water

Water makes up 95 percent of beer's total volume. Its pH balance, as well as the elements the water contains, is critically important to the quality of the end product. When thinking about water, you must consider its source. Public water is often treated with chlorine, which imparts bad flavors to beer. Well water can be high in mineral content. Water that has gone through a water softener will be high in sodium. Water used to make beer in any quantity really should be tested so the brewer knows what's in it and can take measures to correct any potential problems.

Hard water contains a lot of minerals. Soft water contains few minerals. Some minerals improve the flavor of beer, and some make it worse. It's better to start with soft water if you have the option because it is easier to add necessary minerals than it is to take them out. In terms of acid and alkaline for brewing beer, water pH should be around 5.5.

Hops

Relative to water and grain, hops are a minor ingredient of beer by volume, but essential to bitterness and aroma. Brewers use combinations of specific varieties of hops in measured amounts at particular stages during the brewing process to achieve certain results based on the levels of alpha and beta acids and essential oils contained within the hops' lupulin. These substances are released when the hops are added to the boiling brew kettle through a process called isomerization. Home brew recipes call for specific varieties of hops to be added in certain quantities. The recipe tells the brewer at what point the hops should be added and for how long they should be boiled.

Commercial brewers get even more serious. To achieve the results they are looking for, they need to know precise specifics about the hops they are using. Certain hop varieties are known for producing lupulin containing semipredictable levels of acids and essential oils; however, these levels will vary from season to season, region to region, and hop yard to hop yard depending on conditions and handling. Growers test their hops routinely so that they can tell brewers the amount of alpha and beta acids in a given hop harvest, the hop storage index, and (in some cases) a breakdown of the compounds in the hop oil. Alpha and beta acids as well as hop oils oxidize over time. This process can be accelerated by improper handling during harvesting, processing, and storage. The hop storage index reports on the freshness of the hops and gives the brewer an idea of how long the hops will maintain their various component levels (see Chapter 13).

Another factor for the brewer to consider is whether to use whole hop cones, extract, or pellets. Ease of storage and shelf life vary from form to form. In addition the various forms of hops have their advantages and disadvantages during the brewing process in terms of convenience and ease of handling, as well as flavor imparted. Some brewers also use wet hops, which have their own special brewing characteristics.

Hops provide beer's pungent aroma and bitter flavor.

Other variables brewers control to get different results have to do with when the hops are added during the brewing process and how long they spend in the brew kettle before they are removed. Hops are generally added during what is called "the boil." This is when the water left behind after the removal of the spent barley malt is boiled for an hour. Hops to supply bitterness are added at the beginning of the boil and generally stay in the pot for a full hour. Hops intended to provide flavor are added about halfway through the boil. The oils that provide aroma are very volatile and can easily boil away. Therefore aroma hops are added during the last five minutes of the boil. To capture more essential oils, hops can also be added to the fermentation tank or even to the keg containing the finished product. This is called dry-hopping.

Brewing with Wet Hops

Because wet hops have virtually no shelf life, brewers use them fresh off the bine within hours of harvest in seasonal brews that showcase the ephemeral fresh-hop flavor and aroma that appreciative beer drinkers have struggled to describe in numerous ways, including earthy, floral, grassy, vegetative, chlorophyllic, spicy, herbal, and evergreen.

Why do fresh hops produce such different aromas and flavors from dry hops in beer? These differences are created by the levels and types of volatile hop oils contained within the hop cones. As the word "volatile" indicates, these oils have a tendency to easily and quickly transform into vapor—and evaporation begins the minute the hop cones hit the air after harvest. In fact the great rush to dry, process, and package hops after they are harvested is driven by the desire to minimize the loss of the cone's hop oils and acids. Brewing with wet hops bypasses all this and simply introduces the fresh, intensely aromatic hop flowers, sticky with volatile oils, directly into the brew kettle, where the hop's very essence is captured.

Because of the nature of the relatively short hop harvest and extremely short shelf life of fresh hops, wet-hop beer is a truly seasonal product that can only be made during the hop harvest season. In addition, because the freshly picked hops must be used within twenty-four hours of harvest, at its best a wet-hop beer is a truly local product as well. Wet-hop brewing is an opportunity for the local farmer and the local brewer to work in concert to create something both ephemeral and amazing—but it requires planning.

First, a brewer must be interested in working with wet hops. It is harder to brew beer with wet hops, and recipes have to be adjusted. Wet hops are not as bitter by weight as dry hops so a brewer has to add four to six times more wet hops than dry to get the same result. For example, converting to wet hops for a home brew recipe that calls for ½ ounce

Brewers use fresh-picked hops to create a seasonal beverage referred to as wet-hop or fresh-hop ale.

A brewer from Allagash Brewing Company in Maine rubs fresh hop cones between his hands to assess lupulin levels and aroma. Photograph courtesy of The Hop Yard (Gorham, Maine)

(14.2 grams) of dry hops will mean the brewer has to add 2 or 3 ounces (56.7 to 85.1 grams) of wet hops. For a commercial brewer measuring hops in pounds, the additional space needed can be a logistical problem, not only in the brew kettle but in the brewery itself.

But there are bigger issues. In wet-hop brewing timing is everything. The grower needs to let the brewer know as the hops begin to reach maturity, and the brewer has to plan accordingly. Selling wet hops to a brewer is often a great way for a new hop grower, who does not have access to drying and processing equipment, to sell his first crop. If the grower doesn't yet have a harvesting machine, sometimes an enthusiastic brewery will even send their staff to come and handpick the hops. This can be a great opportunity for the brewer and staff to get out of the brewery and reconnect with the agricultural nature of their product. One way or another, once the hop cones are removed from the bines, they have to begin their journey to the brew kettle, and fast. This is a case in which the best product is simply the freshest product. But even the best laid plans can fall apart. If the hops can't be used within the recommended twenty-four hours, some have had luck refrigerating the hops for a few days before they are used. Whatever you do, don't let the hops sit around in bags or boxes at room temperature.

Tim Adams, head brewer at Oxbow Brewing Company in Newcastle, Maine, poses with a handful of wet hops. Photograph courtesy of Andrew Foster

Fresh hop cones being picked into a bin.

Fresh-picked hops, like fresh-cut hay, heat up when they sit. In the case of hops the rising temperature degrades the quality of the hops, accelerating the evaporation of those volatile oils.

If you know it is hop harvest season (late August to mid-September) and you are ready to drink wet-hopped ale, remember that once the fresh-hops beer is brewed it has to sit for two to four weeks before it is ready to drink. So it may not actually be available until later in September or early October. The fresh nature of the product applies to the beer as well, which should be drunk within three months of its brew date. All this adds up to the beer version of Beaujolais nouveau, with its own kind of hype.

Wet-hop brews form the foundation of hop festivals in the hop-growing regions of the northwestern United States. Judging from media coverage and advertising, beer drinkers and brewers alike out there are positively crazed about wet-hop ale and will go to great extremes to make and drink it. Without local hops until recently, the eastern part of the country has been left out of this tradition—but the hop renaissance here means that local breweries are connecting with fledgling hop growers to seek out wet hops. The Hop Yard, in Maine, sold its entire 2014 harvest as wet hops to a handful of Maine breweries. C. H. Evans Brewing Company has brewed beer with wet hops from our farm for two years in a

Brewers at Oxbow Brewing in Newcastle, Maine, prepare a wet-hop ale. Photograph courtesy of Tim Adams

Hop picker Robert Stewart of Troy, New York, is enthusiastic about fresh hops.

row, serving the beer Wet Hop Harvest at the Albany Pump Station in Albany, New York. But brewing with wet hops can be both messy and unpredictable, and some small brewers with limited human resources and production capabilities who are hard-pressed to keep up with demand for their product under the best of circumstances have been understandably reluctant to take the plunge.

Ever-curious home brewers are always anxious to get their hands on wet hops for their beers, and even beer drinkers who are not brewers can experiment by wet-hopping beer that's already been brewed. There is a device brewers use called a Randall that comes in handy for this. It is a plastic cylinder that you can stuff full of hops, wet or dry. One end attaches to the tap and the other to a spigot. The beer is pumped through the hops, absorbing the flavors and aromas, and into the beer glass, ready for consumption.

Another trick for hop lovers is to put hops into a French coffee press and fill it with beer. Let it sit for a while, then press the hops down to the bottom of the container and pour the beer into a glass for a wet-hop-infused drink. An even simpler method of integrating fresh hops, and one that is enjoyed frequently on our farm during harvest, is to fill a glass with beer and drop a few fresh hop flowers into it. Each sip will overwhelm you with hoppiness.

This device, called a Randall, allows beer drinkers to add an extra hoppiness to their beer. Simply stuff the plastic cylinder full of wet or dry hop cones and run tapped beer through the hops on its way into the pint glass.

Adding a few fresh hop cones to a glass of beer provides that fresh hop aroma.

Yeast

Yeast, a single-celled fungus that can reproduce itself, is the secret ingredient that creates the alcohol in beer through that magical process called fermentation. Once the wort suffused with the hops during the boil has cooled, yeast is added, and the mixture is set aside to ferment. The yeast consumes the liquid sugar in the wort, excreting alcohol and carbon dioxide and flavor compounds, while reproducing until all the sugar is gone—at which point fermentation stops.

The two different types of yeast used in brewing produce different styles of beer. Top-fermenting yeast, also called ale yeast, works at a temperature between 60 and 70 degrees Fahrenheit (15.6 and 21.1 degrees Celsius) and takes just a few days to ferment. It produces a rich, fruity beer. Bottom-fermenting yeast, also called lager yeast, works at a temperature between 38 and 50 degrees Fahrenheit (3.3 and 10 degrees Celsius) and takes quite a bit longer to complete fermentation. In fact, in German the word *lager* originally meant "storehouse." Beer brewed with this yeast has a clean, crisp flavor.

Beer-Making Basics

The brewer combines these four ingredients, temperature control, and time to produce beer. Though the process varies for different styles of beer as well as with the scale of production, there are a few basic elements that are constants. The devil remains in the details.

The brewing process starts when the cracked barley malt is mixed with hot water. This is called mashing. Hot water is circulated through the mash to rinse out the sugars in a process called sparging. Once the sugars have been released into the water,

During the brewing process hot water is circulated over the malting barley, picking up sugars and becoming wort.
Photograph courtesy of Megan Reilly

the liquid is separated from the spent barley malt in a process called lautering. The remaining liquid is called wort. The wort is heated in the brew kettle. Once it is boiling the bittering hops are added. The wort boils for one hour.

Halfway through the one-hour boil, flavoring hops are added, with aroma hops being added at the end of the one-hour cycle. The liquid is then chilled to a temperature below 80 degrees Fahrenheit (26.7 degrees Celsius) and siphoned into the fermenter. At this point, before yeast is added and fermentation begins, a hydrometer reading is taken to determine the original specific gravity of the brew, which is essentially a measurement of the percentage of sugars. The yeast is then added. The fermenter is sealed with a stopper holding an airlock and stored in a cool, dark location where within a day it will begin to bubble. The bubbling is caused by the release of carbon dioxide being produced by the yeast as it consumes the sugars. Fermentation will continue for a week or more.

When fermentation is over the bubbling will stop. A sample is taken from the fermenter, and another hydrometer test is performed and compared to the reading taken the day the beer was brewed. If the beer is ready, the final specific gravity will be less than 35 percent of the original reading. This reading can also help you calculate the alcohol by volume of the beer.

Before the beer is bottled it must be primed by adding a fermentable sugar, previously dissolved in boiling water, then cooled, to the beer. The sugar feeds the small population of yeast that remains in the beer postfermentation. The beer is then bottled, capped, and left to sit for another one or two weeks. During this time the remaining yeast will consume the recently added sugar to produce carbon dioxide, which carbonates the beer.

Hops are added to the boiling wort to flavor the beer. Photograph courtesy of Megan Reilly

Hop-centric Styles of Beer

In terms of semantics, at its most basic definition beer is an alcoholic beverage made from fermented grain. There are two primary types of beer—ale and lager. The difference between ale and lager has its origins in the yeast. As described earlier, ale is made from top-fermenting yeast and its fermentation takes place at a warmer temperature and happens relatively quickly. Lager is made with a bottom-fermenting yeast. Its fermentation occurs at lower temperatures and takes longer. Beyond this, ales tend to be made with a broader range of ingredients that are used in larger quantities. Within the ales and lagers there are number of styles with a persona dictated by their geographic origin. Ale styles include American, English, Finnish, German, Irish, Russian, and Scottish. Lager styles include American, Czech, Euro, German, and Japanese.

Hops are used to some extent in all styles of beer. However, for some styles of beer, hops are a signature ingredient. All beers with a high emphasis on hops are considered ales. Within this group of hoppy ales are some styles that have been hijacked by hops. These hop-centric beers are essentially a new American phenomenon and incorporate extremely high-alpha varieties of hops grown, for the most part, in the United States—although some are coming from New Zealand. These beer styles include American IPA, American double/imperial IPA, and American barleywine. This hop-forward trend has its roots on the West Coast, beginning with products such as Sierra Nevada Brewing Company's Pale Ale and Stone Brewing Company's Arrogant Bastard Ale, both out of California.

Interestingly enough, many of today's latest top-rated, uber-hoppy beers are made in the East. Examples include Julius from Tree House Brewing in

A sample of Indian Ladder Stout being poured from a tank at Other Half Brewing Company in Brooklyn, New York.

Sam Richardson of Other Half Brewing Company brewing Indian Ladder Stout.

Massachusetts; Heady Topper, an American double/imperial IPA brewed in Vermont by The Alchemist; Olde School Barleywine, from Dogfish Head Craft Brewery in Delaware; and Hop Showers, another highly rated American IPA, from Other Half Brewing Company in Brooklyn, New York.

What is behind this trend that is producing beers so hoppy they will, in the words of one beer reviewer, "rip your tongue out" (he means in a good way)? Is it because Americans have young, unrefined palates and always want everything to be bigger and better and more outlandish (think Doritos-flavored Mountain Dew)? Possibly. But it is not just us. Nor is it confined to other relatively new beer hotspots such as Australia. Even in the United Kingdom and Europe, home to beer drinkers with refined tastes honed over centuries, the hoppy beer trend is strong. BrewDog, a brewery in Scotland, pitches their signature Punk IPA as "a post Punk apocalyptic mother of an ale . . . subverted with new world hops to create a devastating explosion of flavor." Duvel Moortgat, the

A field of conlon barley sprouting on our farm in early spring.

The same conlon barley as it begins to tassel.

Sam Richardson removing that same barley, now spent after brewing the Indian Ladder Stout, from the Other Half Brewing Company.

brewery based in Belgium we visited earlier in this book, is now producing the annual limited edition Duvel Tripel Hop, a Belgian strong beer brewed with three different varieties of hop—one of which will change each year.

Where is this all going? What role will hops grown in the eastern United States play on the global stage? It is too soon to tell. First we have to demonstrate we can grow hops. What the brewers do with them will be decided by the marketplace.

CHAPTER 15

Beer Recipes

We have a lot of friends who make beer. Whether they are home brewers or brewmasters they are fascinated by the production of the agricultural ingredients that go into their brews and make a point of getting out of the brewery and onto the farm. They love to learn about how hops grow and the array of environmental factors in the field that contribute to the development of the acids and essential oils in the lupulin. They put this knowledge to work when they brew their beer. Below are beer recipes from a few of our brewer friends, each of whom has brewed beer with ingredients grown on our farm.

1835 Albany Ale

BREWER: Craig Gravina
Albany Ale Project
Albany, New York

Craig Gravina is a world-class beer drinker. He is so infatuated with the sudsy stuff he took to writing a blog, *Drink Drank*, about it. Gravina stumbled across the four-hundred-year brewing history of his hometown of Albany, New York, and the long-lost story of Albany Ale. This discovery resulted in an international research effort—The Albany Ale Project (http://www.albanyaleproject.com)—and most recently a book, *Upper Hudson Valley Beer*, coauthored by Alan McLeod.

DESCRIPTION

In the nineteenth century, the brewers of the Upper Hudson Valley were making and exporting hundreds of thousands of barrels of pale, strong double ale to every corner of the United States and abroad. They dubbed these brews "Albany Ale." By the end of the nineteenth century, while the rest of the country was embracing lager, Albany and the Upper Hudson Valley were making three times as much ale as lager. Although Albany Ale's popularity waned in the early twentieth century, the area's ale brewing tradition continued after prohibition and helped to usher in the craft beer revolution.

This brew is typical of the pale, strong double Albany Ales of the early to mid-nineteenth century. This brew uses six-row malt rather than two-row, because six-row was the most common American barley of the nineteenth century—and New York State was one of the largest growers of it at the time. Most of these brews were not sparged; instead they were infused at least twice and often parti-gyled; that is to say, multiple mash runnings were drawn off, hopped and boiled separately, then blended back together—in different proportions—to achieve consistent gravities. Because of the scale of this brew, and for the sake of simplicity, this version employs the "entire butt" method and omits the blending and separate hopping and boils but still includes two mashes. Mash temperatures in nineteenth-century brews were also often quite high.

Hops are always a mystery in historic brews. Although not specified, it's assumed that they were New York–grown hops of the Cluster variety. There is almost no information about how, or when, they were added to boil. Heritage hops, such as Helderbergs, would be ideal for this beer. Additionally, no yeast information is given in the original recipe.

RECIPE

For 5 gallons (18.9 liters) all-grain

Ingredients

17.2 lbs. (7.8 kg.) six-row malt
2 oz. (57 g.) of 6% AAU Cluster hops (90 minutes)
2 oz. (57 g.) of 6% AAU Cluster hops (60 minutes)
2 oz. (57 g.) of 6% AAU Cluster hops (30 minutes)
0.5 to 1 oz (14.2 to 28.4 g.) of 6% AAU Cluster hops dry-hop (optional)
5 oz. (141.8 g.) honey (10 minutes)
1.5 tsp. (7.1 g) salt (10 minutes)
Wyeast 1728 (Scottish Ale) yeast (or any lower attenuating ale yeast)

Specifications

Original Gravity: 1.093
Final Gravity: 1.027
Alcohol by Volume: 8.6%
International Bittering Units: 78
Standard Reference Method: 4
Boil Time: 90 minutes
Pre-boil Volume: 6.5 gallons (24.6 liters)

Directions

Mash your 5-gallon (18.9-liter) batch as usual, aiming for a temperature of 172 degrees Fahrenheit (78 degrees Celsius) and splitting your water volume per mash as you see fit. A one-third/two-third split produces a nice, concentrated wort from the first running.

After an hour, drain off the first running into your kettle, then repeat the same process for the second mash—this time aiming for a mash temperature of 180 degrees Fahrenheit (82 degrees Celsius).

Mash for another hour, then drain off into your kettle and begin your boil.

Boil for 90 minutes, with 2 ounce (57 gram) hop additions at 90, 60, and 30 minutes.

Add 5 ounces (141.8 grams) of honey and 1½ teaspoons (7.1 g) of salt 10 minutes before the end of the boil.

When 90 minutes is up, chill the wort to 65 degrees Fahrenheit (18 degrees Celsius).

Rack to a carboy or fermenting bucket, and pitch the yeast. Ferment for one to two weeks.

Dry-hopping would not have been unusual in the early nineteenth century and was often used as an additional method of preserving beer. If desired, after primary fermentation the beer can be racked to a secondary vessel, ½ to 1 ounce (14.2 to 28.4 grams) of Cluster hops can be added, and the beer can be allowed to sit for another week.

After a week, the beer can be packaged (either bottled or kegged) with a target of 0.75 to 1.3 volumes of CO_2.

Kai Rye

BREWER: David Van Houte
Indian Ladder Farmstead Brewery and Cidery
Altamont, New York

David Van Houte has been brewing beer for ten years and has won awards locally and afar at both sanctioned and unsanctioned beer competitions. His favorite beers are IPAs and malty German lagers. He recently got his PhD in biology from Rensselaer Polytechnic Institute, started a hop farm growing various varieties, and has been involved in starting a malting company in New York's Capital District. Van Houte is the brewer at the Indian Ladder Farmstead Brewery and Cidery.

DESCRIPTION

Rye adds greater character to IPAs and pale ales and pairs well with many of the hop varieties that grow well in the East. I prefer Citra for dry-hopping, but this could easily be replaced by aromatic local hop varieties, as could any of the hops listed for this recipe.

RECIPE

For 5 gallons (18.9 liters) all-grain

Ingredients

12 lbs. (5.4 kg.) Simpsons Golden Promise–78%
2.5 lbs. (1.1 kg.) rye malt–16.33%
2 oz. (57 g.) Weyermann Chocolate Rye–0.82%
11 oz. (312 g.) Caramel 40L–4.9%
1 lb. (0.5 kg.) rice hulls
0.5 oz. (14.2 g.) Columbus (first wort hop)
0.75 oz. (21.3 g.) Magnum (first wort hop)
2 oz. (57 g.) Chinook (whirlpool)
2 oz. (57 g.) Magnum (whirlpool)
1 oz. (28.4 g.) Chinook (five-day dry-hop)
1 oz. Citra (28.4 g.) (five-day dry-hop)
Yeast: California Ale WLP001

Specifications

Original Gravity: 1.068
Final Gravity: 1.016
Alcohol by Volume: 6.88%
International Bittering Units: 81.16
Standard Reference Method: 10.7
Boil Time: 60 minutes
Pre-boil Volume: 5.5 gallons (20.8 liters)

Directions

Bring 4.8 gallons (18.2 liters) of water to 170 degrees Fahrenheit (77 degrees Celsius) and mash all the grains in at 152 degrees Fahrenheit (67 degrees

A beer flight served at a brew pub in Seattle, Washington.

Celsius), mixing thoroughly. Conversion time is 60 minutes. Add 2 gallons (7.6 liters) of 212 degree Fahrenheit (100 degrees Celsius) water to mash out for 10 minutes at 168 degrees Fahrenheit (175 degrees Celsius).

Sparge with 2.8 gallons (10.6 liters) of 170-degree Fahrenheit (76 degrees Celsius) water and collect approximately 5.5 gallons (20.8 liters) of wort, adding first-wort hops to the kettle during lautering.

Boil for 60 minutes, and add whirlpool hop editions at flameout.

Ferment in one fermenter if possible; otherwise limit oxygen exposure after dry-hopping.

Purge with CO_2 all equipment prior to racking.

Ferment in primary for about a week at 68 degrees Fahrenheit (20 degrees Celsius), then for five to seven days in secondary around 72 degrees Fahrenheit (22 degrees Celsius).

Dry-hop during secondary for five days, then begin a three-day cold crash: 55 degrees Fahrenheit (13 degrees Celsius) for one day, 45 degrees Fahrenheit (7 degrees Celsius) for a second day, then 35 degrees Fahrenheit (1.7 degrees Celsius) for the last day.

Bottle or keg with a target of 2.5 volumes of CO_2.

Indian Ladder Farmstead IPA

BREWER: Ryan Demler, brewing operations manager
C. H. Evans Brewing Company
Albany, New York

As with many professional brewers, Demler got his start home brewing. He was an economics student in college and, ever mindful of the impact of drinking on his wallet, dove early on into the world of home brewing. Lacking formal training, Demler was fortunate enough to be offered a job with Olde Saratoga Brewing Company, located in Saratoga Springs, New York. From there he moved on to Cameron's Brewing Company in Oakville, Ontario. In 2012 Demler took over brewing operations at the C. H. Evans Brewing Company's Albany Pump Station, located in Albany, New York, where he produces over fifty different styles of beer annually. Demler enjoys working with local farmers to obtain ingredients as well as with local distilleries for their spent barrels.

DESCRIPTION

This is a truly unique IPA originally brewed with hops and barley grown on Indian Ladder Farms. The beer has a perfumey nose; bitter hoppy bite; and crisp, dry finish. The nose is dominated by floral, almost perfumelike notes from the malt with a bit of spiciness and pineapple from the late hop additions. It pours a very light strawlike color with a tight white head. The beer is very light-bodied and crisp, with a low alcohol content. The perfume nose gives way to a complex hoppy roller-coaster ride of flavor. Starting out with a slightly sweet malt flavor, the spicy Brewer's Gold and resiny Nugget take over the palate until it finishes dry and crisp.

RECIPE

For 5 gallons (18.9 liters)

Ingredients

8.1 lbs. (3.7 kg.) pale malted Newdale
3.5 oz. (99.2 g.) of Brewer's Gold (7.4% AA)
3.5 oz. (99.2 g.) Nugget (11.6% AA)
2 oz. (57 g.) Centennial
2 oz. (57 g.) Helderberg Hops
White Labs California Ale Yeast (WLP001)

Specifications

Original Gravity: 1.050
Final Gravity: 1.010
Alcohol by Volume: 5.2%
International Bittering Units: 48
Standard Reference Method: 3.2
Boil Time: 60 minutes
Pre-Boil Collection: 5.5 gallons (20.8 liters)

Hops are only used in small amounts when brewing beer but provide the essential elements of bitterness and aroma.

Directions

To begin the brew, mix all of your milled grain with 3.5 gallons (13.3 liters) of strike water at approximately 175 degrees Fahrenheit (79 degrees Celsius) to achieve a mash temperature of 152 degrees Fahrenheit (67 degrees Celsius), and mix thoroughly. Add your mash hops of 0.6 ounce (17 grams) of Brewer's Gold, again mixing thoroughly, then rest for 30 minutes. Raise to mash-off temperature of approximately 168 degrees Fahrenheit (76 degrees Celsius), being sure to stir constantly to prevent the bottom of your mash from burning and vorlauf for 20 minutes to clarify wort. Run off into the kettle. During runoff, sparge with 3 gallons (11.4 liters) of water at 171 degrees Fahrenheit (77 degrees Celsius) to collect a total of 5.5 gallons (20.8 liters) of wort in your kettle.

Add your first wort hops of 0.4 ounce (11.3 grams) of Nugget. Bring to a boil, and at 10 minutes remaining add 0.5 ounce (14.2 grams) Brewer's Gold and 0.3 ounce (8.5 grams) Nugget. At the end of boil add 0.4 ounce (11.3 grams) Brewer's Gold. At 15 minutes after the end of boil add 0.8 ounce (22.7 grams) Nugget.

Chill the wort to 67 degrees Fahrenheit (19 degrees Celsius), and transfer to your fermenter, being sure to oxygenate and add your yeast.

When fermentation is nearing an end (a specific gravity reading of 1.016 to 1.020), add your dry hops (2 ounces [57 grams] each of Helderberg, Brewer's Gold, Nugget, and Centennial), and close your fermenter.

Once the beer has achieved terminal gravity, rack into secondary and cold age for at least two weeks. Bottle or keg to 2.4 volumes of CO_2.

Indian Ladder Dry-Hopped Pale Ale

BREWER: Dan Suarez, owner
Suarez Family Brewery
Livingston, New York

Dan Suarez has several years of experience working in commercial breweries in the Northeast, including a three-year stint as the assistant brewer at Hill Farmstead Brewery, located in Greensboro, Vermont. His favorite beers to drink and brew are pilsners, pale ales, and farmhouse-type beers. In 2016 Dan and his wife, Taylor Cocalis Suarez, will open Suarez Family Brewery in Livingston, New York.

DESCRIPTION

This recipe is for a straightforward pale ale showcasing classic resinous American varieties of hops with generous whirlpool and dry-hop additions. Hop substitutions may be made to accommodate different hop varietals.

RECIPE

For 5 gallons (18.9 liters) all-grain

Ingredients

11 lbs. (5 kg.) American 2-Row Malt
(90% of total grist)
9.8 oz. (278 g.) Gambrinus Honey Malt
(5% of total grist)
9.8 oz. (278 g.) American Wheat Malt
(5% of total grist)
0.35 oz. (10 g.) Columbus Pellets (60 minutes)
1.2 oz (35 g.) Centennial Pellets (whirlpool)
1.2 oz. (35 g.) Cascade Pellets for whirlpool
1.2 oz. (35 g.) Columbus hops (dry hop)
1.2 oz (35 g.) Centennial Pellets (dry hop)
Yeast: English Ale Yeast WLP 007

Specifications

Original Gravity: 1.053
Final Gravity: 1.010–1.012
Alcohol by Volume: 5.3–5.6%
International Bittering Units: about 25
Boil Time: 60–90 minutes
Pre-boil Volume: 7.4 gallons (28 liters)
Fermentation Temperature: 68 degrees Fahrenheit
(20 degrees Celsius)

Directions

Mash your grains with 5 gallons (18.9 liters) of hot water to reach a uniform mash temperature of 154 Fahrenheit (68 degrees Celsius). Stir well while you are mashing in to avoid dry clumps of grain and to ensure uniformity of mash temperature. Let mash rest, undisturbed, for at least 60 minutes

Begin the vorlauf process to clarify your mash runnings. Once your wort is running clear and free of grain particulate from the bottom of your mash tun, you may begin collecting it into your kettle. Sparge with 5.28 gallons (20 liters) of water at 170 degrees Fahrenheit (77 degrees Celsius).

Collect 7.4 gallons (28 liters) of wort, and bring to a boil. Boil for 60 minutes, and follow the hopping schedule outlined.

When adding whirlpool hop addition at the end of the boil, allow the whirlpool to rest and settle for 20 minutes before chilling/knocking out. Knock wort out at 66 degrees Fahrenheit (19 degrees Celsius), oxygenate well, and add an appropriate amount of healthy yeast.

Ferment at 68 degrees Fahrenheit (20 degrees Celsius) until you have reached terminal (final) gravity. Once your beer has reached terminal gravity and is free of excessive early fermentation by-products (diacetyl, acetaldehyde, etc.), you may add your dry hops and allow them to steep for 7 days.

Chill beer at least 24 hours before racking. Once beer has been thoroughly chilled, you may rack beer to another sanitary vessel for a period of cold conditioning before packaging. Alternatively, you may package the beer in bottle or keg at this point and target 2.5 volumes of CO_2.

ACKNOWLEDGMENTS

There are many people who helped in the process of writing this book as well as assisting with the work in the hop yard that informed the writing. First and foremost, thanks to Joni Praded and everyone else at Chelsea Green for giving us the opportunity to write this book and helping to make an idea become reality.

For generously imparting their hard-won knowledge about growing hops in the eastern United States we would like to thank Heather Darby, Lily Calderwood, and everyone else at the University of Vermont's Northwest Crops and Soil Program; Roger Rainville of Borderview Farm; and Steve Miller and Jason Townsend of Cornell Cooperative Extension of Madison County, New York. Thanks to the folks at the Carey Center for Global Good's Farm-to-Glass program for all of your support. Thanks also to Tom Gallagher of Cornell Cooperative Extension Albany for helping me learn more about soil fertility. And thanks to author Stan Hieronymus for sharing his knowledge of the ancient history of hop varieties and to Dan Driscoll for giving us his Helderberg Hops.

Thanks to my father, Peter G. Ten Eyck II, and Joe Nuciforo of Indian Ladder Farms for sharing their knowledge of Integrated Pest Management, and thanks to the entire Indian Ladder Farms crew for helping out in all kinds of ways. Thanks to Larry and Kate Fisher at Foothill Hops. Also, thanks to all of the people in Washington and Oregon who took the time to show us how hops are grown and processed on a large commercial scale, including: Paul Matthews, Nicholi Pitra, and David Duham of Hopsteiner; Danny Hallman of Golden Gate Hop Ranch; Gayle Goschie of Goschie Farms; and Michelle Palacios of Crosby Hop Farm. Also thanks to Thomas Shellhammer and Daniel Vollmer of Oregon State University's Brewing Science Laboratory for talking to us about the chemistry of hops in beer.

Thanks to brewers Sam Richardson and Matt Monahan of Other Half Brewing Company; Dan Suarez and Taylor Cocalis Suarez of Suarez Family Brewery; Ryan Demler of C. H. Evans Brewing Company; Craig Gravina of the Albany Ale Project; Garry Brown of Brown's Brewing Company; and our own David Van Houte for sharing their knowledge of hops and brewing.

Particular thanks to our hardworking hop yard crew of retirees and college students, Kathy Meany, Megan Reilly, James Roe, and Ron White.

And last but certainly not least, thanks to all of our friends, family, and neighbors who have helped us establish the Helderberg Hop Farm and the Indian Ladder Farmstead Brewery and Cidery and provided moral support during the writing of this book. You are too numerous to mention, but we are going to try. First of all, thanks to our parents, Charles and Jean Gehring, Mary Jane Fryer, and Peter and Rose Marie Ten Eyck, as well as our son, Wolfgang Gehring. And thanks to all our friends and neighbors, including Chris and Trevan Albright, specializing in equipment repair; Tim Albright, supervisor of the movement of very heavy things; and all the hop pickers and hop yard installers, including the Campanas; the Dowd family, particularly Trish, who read much of this manuscript in draft form and provided invaluable input and encouragement; the Edwards; the Guyers; and the Hotopps; both branches of the King family of Clipp Road; the Jobin-Davises; the McKays; and the Millers; the Morrises of New Jersey, New York, and Vermont; Andy, Erin, and Maddie Pelletier; the Powells; the Raceys; the Sabas; the Stewarts; the Wittmans; Susan Albright, Kelly Albright, Alex Orens; Peter Richards; and Joe Saba. They say it takes a village, and guess what? It's true.

RESOURCES

Books

BOOKS ABOUT HOPS

There is not an abundance of books about hops, but below are a few good ones that give great information on the history of hop production, the industry today, and the science behind why hops make beer taste and smell good.

Barth, Henrich Joh, Christiane Klinke, and Claus Schmidt. *The Hop Atlas: The History and Geography of the Cultivated Plant.* 1st English ed. Nuremberg, Germany: Joh. Barth & Sohn, 1994.

Hieronymus, Stan. *For the Love of Hops: The Practical Guide to Aroma, Bitterness, and the Culture of Hops*, Brewing Elements. Boulder, CO: Brewers Publications, 2012.

Tomlan, Michael. *Tinged with Gold: Hop Culture in the United States.* 1st ed. Athens: University of Georgia Press, 1992.

BOOKS ABOUT GROWING HOPS

We consulted a number of texts about growing hops to ferret out the best information for small-scale, sustainable hop growers. Not all of the following texts carry the same advice that we provide here since they are mostly geared toward large commercial growers in the Northwest, but they are essential resources for hop growers of all sizes and geographies.

Technical References

Gent, David H., James D. Barbour, Amy J. Dreves, David G. James, Robert Parker, and Douglas B. Walsh, eds. *Field Guide for Integrated Pest Management in Hops.* Moxie, WA: Cooperative publication produced by Oregon State University, University of Idaho, U.S. Department of Agriculture–Agricultural Research Service, and Washington State University, 2010. http://ipm.wsu.edu/field /pdf/hophandbook2009.pdf.

Hopunion. *The Hop Variety Handbook.* Yakima, WA: Hopunion, 2013.

Mahaffee, Walter F., Sarah J. Pethybridge, and David H. Ghent. *Compendium of Hop Diseases and Pests*, Disease compendium series. St. Paul: APS Press, 2009.

Weigle, Timothy. *2014 Cornell Integrated Hops Production Guide.* Ithaca, NY: Cornell University Cooperative Extension, 2014.

Beginner Guides

There is not a lot of light reading when it comes to the subject of growing hops, but the books below are fun and written with the garden-variety hop grower in mind.

Beach, David. *Homegrown Hops: An Illustrated How-to-Do-It Manual.* 3rd ed. Junction City, OR: David Beach, 1988.

Fisher, Joe, and Dennis Fisher. *The Homebrewer's Garden: How to Easily Grow, Prepare, and Use Your Own Hops, Malts, Brewing Herbs.* Pownal, VT: Storey Books, 1998.

Peragine, Jr., John N. *The Complete Guide to Growing Your Own Hops, Malts, and Brewing Herbs: Everything You Need to Know Explained Simply.* Ocala, FL: Atlantic Publishing Group, 2011.

BOOKS ABOUT BEER

These books provide a wide look at beer from its days as a medicinal beverage to today's elevated realm of craft beer.

Acitelli, Tom. *The Audacity of Hops: The History of America's Craft Beer Revolution.* Chicago: Chicago Review Press, 2013.

Buhner, Stephen Harrod. *Sacred and Herbal Healing Beers: The Secrets of Ancient Fermentation.* Boulder: Siris Books, 1998.

Steele, Mitch. *IPA: Brewing Techniques, Recipes, and the Evolution of India Pale Ale.* Boulder: Brewers Publications, 2012.

BOOKS ABOUT HOW TO BREW BEER AT HOME

Below are two of Dieter's favorite books about home brewing.

Koch, Greg and Matt Allyn, *The Brewer's Apprentice: An Insider's Guide to the Art and Craft of Beer Brewing, Taught by the Masters.* Beverly, MA: Quarry Books, 2011.

Palmer, John. *How to Brew: Everything You Need to Know to Brew Beer Right the First Time.* Boulder: Brewers Publications, 1999.

Additional Important Resources for Hop Growers

For growers in areas new to hop production, such as many regions in the eastern United States, many resources can prove unhelpful, if not misleading. They are either geared for large-scale commercial production in the Northwest, which has unique growing conditions, or (if covering small-scale hops production) they are often isolated to garden-variety growing with few details on siting, trellising, cultivation, or harvesting. We have relied on our own experience as hops growers as well as a wide body of existing and emerging research from a number of growers and researchers. We have found the materials put out by the University of Vermont and Cornell Extension agencies to be invaluable; both programs focus on developing the knowledge necessary to fuel the hops renaissance in the East and have many helpful tools for small-scale growers.

BOOKS

Kneen, Rebecca. *Small Scale and Organic Hops Production.* British Columbia: Left Fields. http://www.crannogales.com/HopsManual.pdf.

ORGANIZATIONS

Steve Miller
Cornell Hops Specialist
Cornell Cooperative Extension Madison County Hops Program
100 Eaton Street
Morrisville, NY 13408
Phone: 315-684-3001, ext. 127
Fax: 315-684-9290
http://madisoncountycce.org/agriculture/hops-program

Heather Darby
Agronomy Specialist
University of Vermont, Northwest Crops and Soil Program
278 S. Main Street
St. Albans, VT 05478
Phone: 802-524-6501, ext. 437
Email: heather.darby@uvm.edu

Northeast Hop Alliance
http://www.northeasthopalliance.org/

Great Lakes Hops
4135 80th Avenue
Zeeland, MI 49464
Phone: 616-875-7416
E-mail: greatlakeshops@gmail.com
http://www.greatlakeshops.com

Michigan State University Extension
http://hops.msu.edu/
As hop production returns to the Eastern United States regional organizations are springing up to help hop growers. Some of these organizations are more established than others with official websites; others are little more than a Facebook group. We have provided available information here to help new growers connect with existing networks in their region.

Maryland Hop Growers Association
Chapter of the Northeast Hop Alliance
https://www.facebook.com/pages/
Maryland-Hop-Growers-Association
/208992835809884?sk

Michigan Hop Alliance
10065 East Island Court
Traverse City, Michigan 49684
Phone: 616-403-6880
http://michiganhopalliance.com/

North Carolina Hops Project
North Carolina State University
Cooperative Extension
Campus Box 7602
Raleigh, NC 27695-7602
Phone: 919-515-2813
http://www.ces.ncsu.edu/fletcher/programs
/nchops/

Ohio State University Hops Research
Ohio State University
College of Food, Agricultural and
Environmental Sciences
South Centers
1864 Shyville Road
Piketon, OH 45661
Phone: 740-289-2071
http://southcenters.osu.edu/horticulture
/other-specialties/hops

Old Dominion Hops Cooperative
PO Box 261
Lyndhurst, VA 22952
http://www.olddominionhops.com/

University of Vermont Extension N.W. Crops and Soils Program
United States Department of Agriculture
Natural Resources Conservation Service
Web Soil Survey
http://websoilsurvey.sc.egov.usda.gov/App
/HomePage.htm

Information on Soil Fertility for Growing Hops

Below are some very useful Cooperative Extension fact sheets that will help you understand the important and complicated issue of nitrogen and hop production.

Gingrich, Gail, John Hart, and Neil Christensen. *Fertilizer Guide: Hops*. Corvallis: Oregon State University Extension Service, 2000. http://buffalo.uwex .edu/files/2011/01/fertilizer-guide.pdf.

Sirrine, Rob. *Hops Fertility Part I: Proper Nitrogen Application and Timing Is Crucial for Maximizing Hop Yields*. East Lansing: Michigan State University Extension, 2014. http://msue.anr.msu.edu/news/hops_fertility_part_i.

Darby, Heather. *Fertility Guideline for Hops in the Northeast.* St. Albans: University of Vermont Extension, 2011. http://www.uvm.edu/extension/cropsoil/wp-content/uploads/HopFertilityManagementNE.pdf.

Resources for Purchasing Rhizomes, Vegetative Cuttings, and Plants

RHIZOMES

Although there is no source for certified virus-free rhizomes, we have consulted with experts in the field of hop production to put together the listing below of reliable sources of quality hop rhizomes.

Crosby Hop Farm
8648 Crosby Road NE
Woodburn, OR 97071
Phone: 503-982-5166
Fax: 503-981-2141
http://crosbyhops.com/product/hop-rhizomes/

Foothill Hops
5024 Bear Path Road
Munnsville, NY 13409
Phone: 315-495-2451
http://www.foothillhops.com

Freshops
36180 Kings Valley Highway
Philomath, OR 97370
Phone: 800-460-6925
Fax: 541-929-2702
http://freshops.com/

Hops Direct
686 Green Valley Road
Mabton, WA 98935
Phone: 888-972-3616
Fax: 509-837-6577
E-mail: info@hopsdirect.com
http://www.hopsdirect.com/rhizomes/

Lone Oak Hop Farm
6021 Deconinck Road NE
Woodburn, OR 97071
Phone: 503-932-3887
E-mail: hopfever@gmail.com
http://www.loneoakhop.com/

PLANTS AND VEGETATIVE CUTTINGS

The list below provides sources for vegetative cuttings from certified virus-free hop plants.

Bundschuh's Greenhouses
1033 Victor Road
Macedon, NY 14502
Phone: 315-986-8872
http://www.bundschuhsgreenhouses.com

Cornell Cooperative Extension of Madison County Potted Hop Plant Sale
Cornell Cooperative Extension Madison County Hops Program
100 Eaton Street
Morrisville, NY 13408
Phone: 315-684-3001
Fax: 315-684-9290
http://madisoncountycce.org/agriculture/hops-program/potted-hop-plants-for-sale

National Clean Plant Network, Clean Plant Center Northwest
Hamilton Hall
24106 N. Bunn Road
Prosser, WA 99350-8694
Phone: 509-786-9242
E-mail: healthy.plants@wsu.edu
http://healthyplants.wsu.edu/hop-program-at-cpcnw/purchasing-hop-material/

Zerrillo Greenhouses Inc.
7581 East Taft Road
East Syracuse, NY 13057
Phone: 315-656-8466

USEFUL INFORMATION ON THE PROPAGATION OF CUTTINGS

Hop Plant Propagation Tips from the Clean Plant Center Northwest
http://healthyplants.wsu.edu/wp-content/uploads/2013/05/Hop-Prop-Tips-R6.pdf

Laboratories

Different labs perform different types of testing useful to hop growers. There are many labs around the country. We have compiled the list below of laboratories we have worked with.

SOIL AND PETIOLE TESTING

Agro-One Soils Laboratory
730 Warren Road
Ithaca, NY 14850
Phone: 800-344-2697
Fax: 607-257-6808
E-mail: soil@dairyone.com
http://dairyone.com/analytical-services/agronomy-services/soil-testing/

COMPOST AND PRE-SIDEDRESS NITRATE TESTING

Cornell Nutrient Analysis Laboratory
804 Bradfield Hall
Ithaca, NY 14853
Phone: 607-255-4540
Fax: 607-255-7656
E-mail: soiltest@cornell.edu
http://cnal.cals.cornell.edu/

HOPS ANALYSIS

Alpha Analytics
203 Division Street
Yakima, WA 98902
Phone: 800-952-4873
Fax: 509-895-7968
E-mail: info@alphaanalyticstesting.com
https://www.alphaanalyticstesting.com/

University of Vermont Cereal Grain Testing Lab
Jeffords Hall, Room 244
63 Carrington Dr.
Burlington, VT 05405
Phone: 802-656-5392
cropsoilvt@gmail.com or erica.cummings@uvm.edu
www.uvm.edu/extension/cropsoil/hops
When submitting hops for testing write "Hops" on the outside of the package.

Insecticides

Below is a link to an extremely handy chart of insecticides approved for use on hops in several northeastern states. The chart specifies those insecticides that can be used on certified organic farms.

University of Vermont Table of Approved Insecticides for Massachusetts, New York, and Vermont
Table 1: Approved Insecticides on Hops in MA, NY, and VT for 2012
(scroll to bottom of below PDF to see chart)
http://www.uvm.edu/extension/cropsoil/wp-content/uploads/Japanese_beetle_in_hops.pdf

Moisture Calculator

When timing your hop harvest, calculating the percentage of dry matter in your hops is crucial. Below is a link to a handy calculator.

University of Vermont Hop Harvest Moisture Calculator
http://www.uvm.edu/extension/agriculture/engineering/?Page=hopscalc.html

Equipment

One of the biggest challenges for small-scale commercial hop growers is finding the right-size processing infrastructure. Below is a list of suppliers we have worked with or learned about through our research for this book.

MECHANICAL HOP HARVESTERS

Tom Frazer
Dauenhauer Manufacturing Company
PO Box 6764
Ketchum, ID 83340
Phone: 208-928-7411
E-mail: tfrazer@dmfg.com
http://www.dmfg.com

University of Vermont Mobile Hop Harvester
Christopher W. Callahan, PE
Agricultural Engineering
University of Vermont Extension
Rutland Office
Howe Center Business Park
1 Scale Avenue, Ste. 55, Rutland, VT 05701-4457
Phone: 802-773 3349 ext. 277
E-mail: chris.callahan@uvm.edu
Factsheet: http://www.uvm.edu/extension/cropsoil/wp-content/uploads/Hops-Harvester-Factsheet.pdf
Design: http://www.uvm.edu/extension/cropsoil/wikis
Video: https://www.youtube.com/watch?v=2iZIkdozeXo

John Condzella
Condzella Hops
Broker for Wolf Hop Harvesters, Drying and Baling Equipment
6233 North Country Road
Wading River, NY 11792
Phone: 631-461-3841
E mail: condzellasfarm@gmail.com

US Hop Source
Wolf harvesting, drying, and baling equipment
7500 E. Dartmouth Avenue
Denver, CO 80231
Phone: 970-240-9056
E-mail: ushopsource@gmail.com

DRYING AND BALING EQUIPMENT

Hop Dryers

Steenland Manufacturing
54400 State Highway 30
Roxbury, NY 12474
Phone: 607-326-7707
http://steenlandmanufacturing.com/

University of Vermont Modular Hop Oast
Factsheet: http://www.uvm.edu/extension/cropsoil/wp-content/uploads/Hops-Oast-Factsheet1.pdf
Design: http://www.uvm.edu/extension/cropsoil/wikis

Moisture Meter

Moist-Vu DL4000
Reid Instruments
Phone: 509-876-2703
www.reidinstruments.com

Hop Baler

University of Vermont Log Splitter Converted to Hop Baler
http://www.uvm.edu/extension/cropsoil/wp-content/uploads/Baler_Instructions_5_4_12.pdf

SOURCES FOR IRRIGATION SUPPLIES

Belle Terre Irrigation
8142 Champlin Road
Sodus, NY 14551
Phone: 315-483-6155
Fax: 315-483-4064

E-mail: info@dripsupply.com
http://www.dripsupply.com

Brookdale Fruit Farm
PO Box 389
41 Broad Street
Hollis, NH 03049
Phone: 603-465-2240 ext. 3
Fax: 603-465-3754
E-mail: tractortrv@aol.com
http://www.brookdalefruitfarm.com/Irrigation/

SOURCES FOR HOP YARD SUPPLIES

Fehr Bros Industries, Inc.
895 Kings Highway
Saugerties, NY 12477
Phone: 800-431-3095
Fax: 888-352-1790
E-mail: info@fehr.com
http://www.fehr.com

Growers Supply
202 South Division Street
Toppenish, WA 98948
Phone: 509-865-3731
E-mail: info@growerssupply.net
http://www.growerssupply.net

Schmidt Farm
415 County Road 8
Farmington, NY 14425
Phone: 585-869-9641
E-mail: sschmidt@schmidthops.com
http://schmidthops.com/
Note the useful Cable Chart from Schmidt Farm:
http://www.schmidthops.com/hop_yard_cable

Hop Pelletizing Services

Foothill Hops
5024 Bear Path Road
Munnsville, NY 13409
Phone: 315-495-2451
http://www.foothillhops.com

Northern Eagle Hop Pelletizing Company
7 Railroad Avenue
Oneonta, NY 13820
Phone: 607-432-0400
E-mail: info@hoppelletizing.com
http://www.hoppelletizing.com

Whipple Brothers Farms
2348 Norway Road
Kendall, NY 14476
Phone: 585-350-9707
E-mail: justin.whipple@gmail.com
http://www.whipplebrothersfarms.com

INDEX

Page numbers in *italics* refer to photographs and figures; page numbers followed by *t* refer to tables.

ABOUT THE AUTHORS

Dietrich Gehring

John Carl D'Annibale

LAURA TEN EYCK owns and operates Indian Ladder Farmstead Brewery and Cidery with her husband, Dietrich Gehring. Their commercial hop yard is located on Indian Ladder Farms, an extensive pick-your-own orchard with a local foods grocery, bakery, café, and retail shop in upstate New York that has been in Ten Eyck's family for four generations, and where she and Gehring live, garden extensively, and raise sheep, dairy goats, and chickens. Ten Eyck is senior manager of New York projects and outreach at American Farmland Trust, where she advocates for national and regional farmland conservation. She was previously a freelance journalist. For more information, see www.ilfbc.com.

DIETRICH GEHRING is a small-scale commercial hop grower, professional photographer, home brewer, and co-owner, with his wife, Laura Ten Eyck, of Indian Ladder Farmstead Brewery and Cidery. He grew hops on a home scale for twenty-five years before starting their commercial hop yard. Gehring grew up working on his grandparents' dairy farm, attended the New England School of Art and Design, and later worked as a photo editor for Workman and other publishers. His long love of hops and brewing began decades ago, when he was sales manager for Newman's Albany Brewing Company, one of the first craft breweries in the United States. His photography has appeared in numerous galleries and magazines. For more information, visit www.dietrichgehring.com.